Aquaculture

NIPA® GENX ELECTRONIC RESOURCES & SOLUTIONS P. LTD.
New Delhi-110 034

About the Editors

Durgesh Kumar Verma completed B.F.Sc. & M.F.Sc. in Fisheries Resource Management from the Acharya Narendra Deva University of Agriculture and Technology, Kumarganj Ayodhya, India. He has authored/published several research and review papers, popular articles, chapters, and books. He is an active and dedicated member of several professional societies mainly engaged in fisheries research and development in the country. He receives several accolades and awards such as Junior Researcher Award, Young Scientist Award, Best Paper Presentation awards, etc. at national and International meetings. Presently he is working as a young researcher at ICAR-CIFRI, Prayagraj, Uttar Pradesh.

Dr. Sanjay Kumar Gupta, M.F.Sc., Ph.D., has published many research and review papers in both national and international peer-reviewed journals. He has written/edited 15 books and 35 book chapters. He visited Australia, Belgium, Netherlands, France, Thailand, Indonesia and Malaysia. Dr. Gupta has presented papers in diverse national and international symposia, workshops, seminars and conferences. He is the recipient of a prestigious endeavour postdoc fellow by the Australian Government (2018). He has been honored with several academic and professional awards for his outstanding contribution in the field of aquaculture and fisheries. Presently, he is working as a Senior Scientist at ICAR-Indian Institute of Agricultural Biotechnology, Ranchi, Jharkhand, India.

Dr. Ashfauq Farooq Aga is a distinguished scientist with B.FSc, M.FSc. & Ph.D. specializing in aquaculture, affiliated with the Division of Fisheries at the Faculty of Fisheries SKUAST-K. He has made significant contributions to the field through his research and academic endeavours'. Aga has served as the editor of several notable books in the domain of fisheries, showcasing his expertise and leadership in compiling and disseminating knowledge in this specialized area of study. He is recipient of best scientist award by PFGF Jammu & Kashmir. He has participated and presented papers in diverse conferences, workshops, and seminars related to aquaculture R&D and education. His biodata reflects his academic achievements, research interests, and leadership in the field of fisheries science.

Dr. Sudeshna Sarker, HOD, Aquatic Animal Health Management, CoF, Kishanganj, BASU is an experienced researcher with a demonstrated history of working in the field of Aquatic Animal Health. Dr. Sarker secured “Dr. N.R. MENON Award” for the ‘Best Indian M.F.Sc Thesis-2016’ by PFGF, India, and is recipient of International Travel Support (ITS) by SERB, GOI for participating in the 5th International conference on "FLAVOBACTERIUM 2018, Japan. She has published research papers, several book chapters, popular articles, and edited a book. She has earlier worked as an Assistant Professor in the School of Agriculture and Allied Sciences, The Neotia University. Currently, Dr. Sarker is posted as Assistant Professor-cum-Junior Scientist in the Dept. of Aquatic Animal Health Management, under Bihar Animal Sciences University.

Aquaculture
Prospects, Opportunities and Challenges

Durgesh Kumar Verma
Young Researcher
ICAR-CIFRI, Prayagraj
Uttar Pradesh, India

Sanjay Kumar Gupta
Senior Scientist
ICAR-Indian Institute of Agricultural Biotechnology
Ranchi, Jharkhand, India

Ashfauq Farooq Aga
Assistant Professor
Division of Fisheries
Sher-e-Kashmir University of Agricultural Sciences and Technology
Shalimar, Srinagar, Jammu and Kashimir, India

Sudeshna Sarker
Assistant Professor and Head
Department of Aquatic Animal Health Management
College of Fisheries, Bihar Animal Sciences University
Bihar, India

NIPA® GENX ELECTRONIC RESOURCES & SOLUTIONS P. LTD.
New Delhi-110 034

NIPA® GENX ELECTRONIC RESOURCES & SOLUTIONS P. LTD.

101,103, Vikas Surya Plaza, CU Block
L.S.C. Market, Pitam Pura, New Delhi-110 034
Ph : +91-11-43860225, Mob.: +91 9717133558, 9540816132
E-mail: newindiapublishingagency@gmail.com
Website: www.nipaersources.com

Print ISBN: 978-93-58878-15-8
ebook ISBN: 978-93-58879-27-8

Composed and Designed by NIPA®.

Preface

"Aquaculture: Prospects, Opportunities, and Challenges," an in-depth exploration of the ever-changing realm of aquaculture and fisheries. This comprehensive volume aims to equip readers with a solid grasp of the industry's historical background, present-day situation, and potential future developments. It offers crucial perspectives on the anticipated opportunities and obstacles that lie ahead in this field.

Our journey begins with a **Historical Overview of Aquaculture and Fisheries and its Evolution**, tracing the development of these crucial industries from their origins to their modern manifestations. This historical perspective sets the stage for understanding the profound transformations that have shaped today's aquaculture and fisheries landscapes.

The book then shifts to an analysis of **The Current Status of Aquaculture and Fisheries: Challenges and Opportunities**, and exploring the present state of the industry. This section highlights both the obstacles that need to be addressed and promising opportunities for growth and innovation.

Looking forward, we delve into **The Future of Aquaculture: Prospects and Challenges**, where we discuss emerging trends and potential developments that could redefine the sector. **Aquaculture and Climate Change** examine how climate change impacts aquaculture practices and strategies that can be employed to adapt to these changes.

The environmental impact of aquaculture is a significant concern in **Aquaculture and Its Impact on The Environment**. This section discusses the effects of aquaculture on ecosystems and explores approaches to mitigate the negative impacts while promoting sustainability.

Bioprospecting in Fisheries and Aquaculture highlights the exploration of biological resources and their potential applications in aquaculture, thereby offering new avenues for innovation and development. **Minimal Water Usage Systems in Aquaculture** presents cutting-edge methods that aim to conserve water resources while maintaining productivity.

As technology continues to advance, **Precision Fisheries: The Future of Farming** explores how precision tools and techniques are transforming fish farming and enhancing efficiency and sustainability. Similarly, **Recent**

Trends and Innovative Technology in Fishing Technology covers the latest technological breakthroughs in fishing practices, revealing their implications for the industry.

Sustainable Aquaculture Practices and Approaches provide a detailed overview of methods and strategies designed to promote long-term ecological balance and productivity. **Organic Fish Farming: Certification for Sustainable Income and Future Importance** discuss the growing importance of organic certification and its role in ensuring sustainable income for fish farmers.

The book also covers economic aspects, with **Fish Hatchery: Good Business for Doubling Income and Aquaculture and Technology for Significant Income Generation of Fish Farmers** discussing how hatcheries and technological advancements can boost profitability in fish farming. **Integrated Fish Farming as Future Importance and Increasing Possibilities of Income to Fish Farmers** explores how integrating various farming practices can enhance financial returns and operational efficiency.

The Role of Government Policy in Shaping the Future of Fisheries and Aquaculture is to examine how policy decisions influence the development and sustainability of the sector. Finally, **Aquaculture Finance and Rural Development** explore the financial aspects of aquaculture and their role in supporting rural development and economic growth.

This book aims to serve as a vital resource for researchers, practitioners, policymakers, and anyone interested in aquaculture and fisheries. By presenting a comprehensive view of the sector's past, present, and future, we hope to inspire and inform efforts to address these challenges and seize the opportunities that lie ahead.

Thank you for joining us in this journey through the prospects, opportunities, and challenges of aquaculture. We invite you to explore the insights and innovations presented on these pages and contribute to the ongoing evolution of this vital industry.

Sincerely,

Editors

Contents

List of Colour Plates

Abbreviations

ABNJ	Areas Beyond National Jurisdiction
ACD	Agricultural Credit Department
AD	Anno Domini
AFGPs	Antifreeze Glycopeptides
AFPs	Asian Federation for Pharmaceutical Sciences
AI	Artificial Intelligence
AICRP	All India Coordinated Research Project
AIS	Automatic Identification Systems
ASC	Aquaculture Stewardship Council
BC	Before Christ
BFT	Biofloc Technology
BMPs	Best Management Practices
BRDs	Bycatch Reduction Devices
CAA	Coastal Aquaculture Authority
CIFA	Central Institute of Freshwater Aquaculture
CIFE	Central Institute of Fisheries Education
CIFRI	Central Inland Fisheries Research Institute
CIFT	Central Institute of Fisheries Technology
CO_2	Carbon Dioxide
CSA	Community-Supported agriculture
DFT	Deep Flow Technique
DSS	Decision Support Systems
EAFM	Economic Administration and Financial Management
EEZ	Exclusive Economic Zone
EwE	Ecopath with Ecosim
FAO	Food and Agriculture Organization
FCH	Floating Capillary Hydroponics
FDA	Food and Drug Administration
FIDF	Fisheries and Aquaculture Infrastructure Development Fund
FIDF	Fisheries Infrastructure Development Fund
FM	Fishmeal

FVC	Forced Vital Capacity
FY	Financial Year
GAA	Global Aquaculture Alliance
GAMs	Generalized Additive Models
GDP	Gross Domestic Product
GIFT	Genetically Improved Farmed Tilapia
GIS	Geographic Information System
GMDSS	Global Maritime Distress and Safety System
GPS	Global Positioning System
GS	Genomic Selection
HIPO	Hierarchical Input Process Output
HIV	Human immunodeficiency virus
ICES	International Council for the Exploration of the Sea
IFDS	Integrated Fisheries Development Scheme
IMTA	Integrated Multi-Trophic Aquaculture
IoT	Internet of Things
IPCC	Intergovernmental Panel on Climate Change
IUU	Illicit, Unreported, and Unregulated
JFE-SSD	Juvenile Fish Excluder cum Shrimp Sorting Device
JFE-SSDJ	Juvenile Fish Excluder Cum Shrimp Sorting Device
JTED	Juvenile and Trash Excluder Device
KVAFSU	Karnataka Veterinary, Animal and Fisheries Science University
MAS	Marker-Assisted Selection
MBBR	Moving Bed Biofilm Reactor
MFRAs	The Marine Fishing Regulation Acts
MPEDA	Marine Products Export Development Authority
MSP	Marine Spatial Planning
NABARD	National Bank for Agriculture and Rural Development
NDSP	New Deep Sea Fishing Policy
NFAP	National Fisheries and Aquaculture Policy
NFDB	National Fisheries Development Board
NFT	Nutrient Film Technique
NGOs	Non-Governmental Organizations
NOAA	National Oceanic and Atmospheric Administration
NPOA-MCS	National Plan of Action for Conservation and Management of Sharks
PA	Polyamide
PE	Polyethylene
PES	Polyester

PFZ	Potential Fishing Zone
PMMSY	Pradhan Mantri Matsya Sampada Yojana
PP	Polypropylene
PVAA	Polyvinyl Alcohol
PVC	Polyvinyl Chloride
PVD	Polyvinylidene Chloride
RAS	Recirculating Aquaculture Systems
RBI	Reserve Bank of India's
RKVY	Rashtriya Krishi Vikas Yojana
RNA	Ribonucleic Acid
SDG	Sustainable Development Goal
SDM	Species Distribution Modeling
UHMWPE	Ultrahigh Molecular Weight Polyethylene
UV	Ultraviolet
VMS	Vessel Monitoring Systems
VTVD	Variable Thrust Vector Devices
WWF	World Wildlife Fund

1

Historical Overview of Aquaculture and Fisheries and its Evolution

Suman, Pratibha Gaurav and Gautam Geeta J.

Department of Zoology, Mahila Mahavidyalaya, BHU, Varanasi-221005
Uttar Pradesh, India

Abstract

Aquaculture is associated with the rearing, breeding, and harvesting of aquatic organisms in cages, tanks, ponds, or other simulated systems, and can be practiced in fresh water, brackish water, or marine water. The practice of cultivating aquatic species in a systematic manner dates back to before 1000 BC. Theories such as oxbow, catch-and-hold, concentration, and trap-and-cropexplain how the transition from hunting to farming may have occurred along with the advancement of civilization. Along with rearing the aquatic species for nourishment, advanced techniques have led to the enhancement of economic value through the inclusion of pearl oyster farming. India is among the largest producers of fisheries and has become the world's second largest producer of aquaculture fish. Globally, aquaculture diversity has reached a maximum value of 428 ± 11 species, of which 29 ± 1 will produce 80% of the global annual production. The total global aquaculture production has increased enormously, from 20 million tons in 1992 to more than 120 million tons in 2020. Fisheries and aquaculture have become essential foods providing commodities worldwide and an important means of nourishment for the poor. A temporal and geographical analysis of global aquaculture diversity reveals an original and hopefully enlightening vision of modern aquaculture and its perspectives.

Key words: *Aquaculture, History, Evolution theories, Indian aquaculture, Aquaculture diversity*

1. Introduction

The farm of aquatic plants and animals is several thousand years old. Aquaculture involves the cultivation of aquatic organisms, including fish, molluscs, crustaceans, and plants in controlled or semi-controlled environments. To rear and increase the number of aquatic species for various purposes, such as food production, recreation, and conservation. Aquaculture is associated with the rearing, breeding, and harvesting of aquatic organisms in cages, tanks, ponds, or other simulated systems. Depending on the specific species being cultivated, aquaculture can be practiced in freshwater, brackish water, or marine water (Barnabe *et al.*, 2005).

According to the Food and Agriculture Organization (FAO), "aquaculture is the farming of aquatic organisms such as fish, mollusks, crustaceans, and aquatic plants." It involves interventions in the rearing process to enhance production, such as regular stocking, feeding, or protection from predators (Tacon *et al.*, 2003).

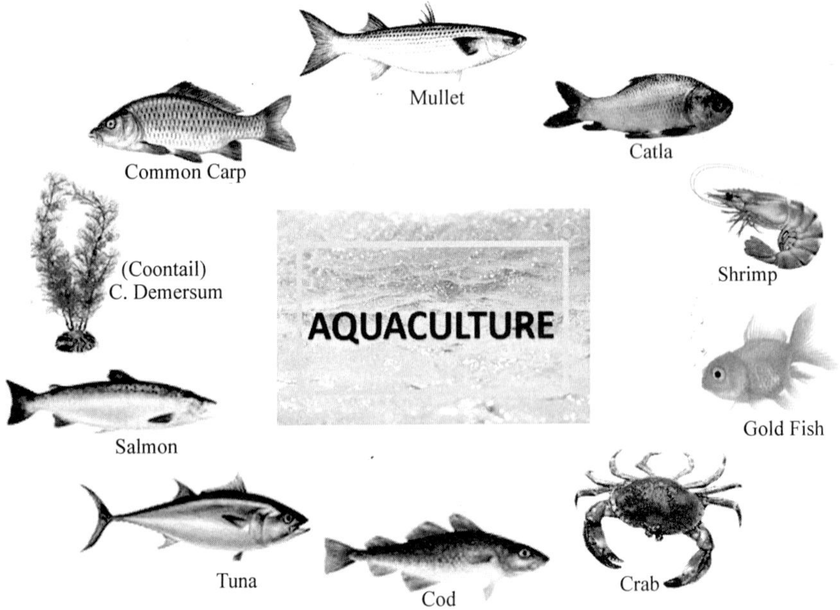

Fig. 1. Image showing different types of organisms cultivated in aquaculture. (Redrawn from- https://www.fisheriesindia.com/2021/04/types-of-aquaculture-and-recent.html)

Aquaculture typically involves the enclosure of a species in a secure system under conditions in which it can thrive; the choice of aquaculture system depends on factors such as the target species, available resources, environmental conditions, and economic considerations. Sustainable aquaculture practices aim to minimize the environmental impact, optimize resource use, and ensure the well-being of cultivated species. Ongoing research and technological developments are continuing to improve the efficiency and sustainability of various aquaculture systems.

Some aquaculture operations combine the elements of different systems to benefit from their respective advantages. For example, integrating a Recirculatory Aquaculture System with ponds or raceways can optimize production efficiency and environmental sustainability.

Aquaculture can be classified on the basis of different factors:-

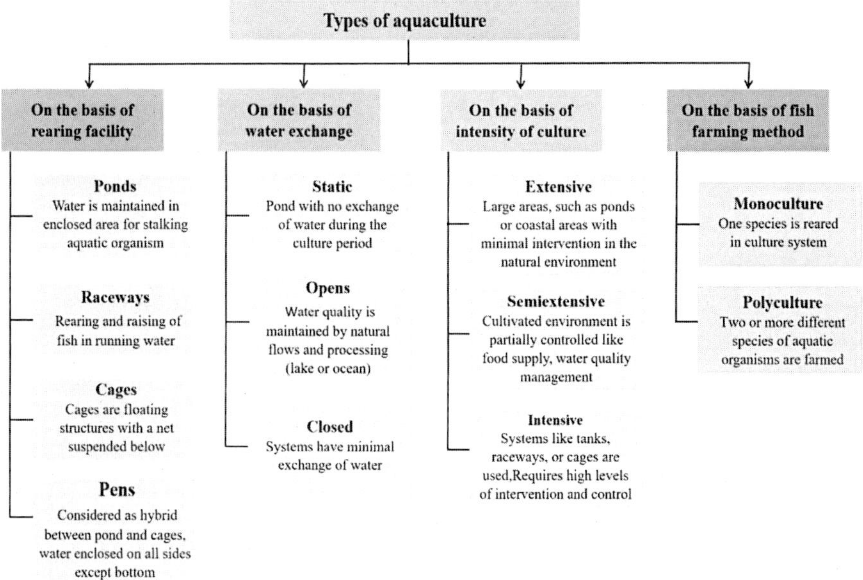

Fig. 2. Image showing types of aquaculture on the basis of rearing facility, water exchange, intensity of culture and fish farming method. (https://www.fisheriesindia.com/2021/04/types-of-aquaculture-and-recent.html)

Aquaculture production has been reported in three distinct culture environments: fresh water, brackish water, and marine water, and freshwater, brackish water, and marine water environments, indicating a diverse range of aquaculture practices. Each environment has a specific set of species that can be cultured successfully, and agriculturists adapt their methods to the characteristics of the water in which they operate. This diversity allows for a wide variety of aquatic products to be produced through aquaculture to meet the demands of global markets (Tacon *et al.*, 2020).

The choice of aquaculture system depends on factors such as target species, available resources, environmental conditions, and economic considerations. Sustainable aquaculture practices aim to minimize the environmental impact, optimize resource use, and ensure the well-being of cultivated species.

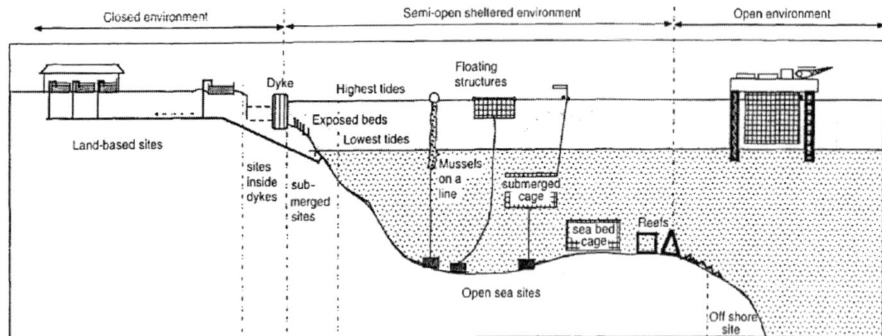

Fig. 3. Sites used for sea water culture including closed environment, semi-open sheltered environment and open environment system (Guilbert *et al.*, 2005)

2. Aquaculture diversity-

Aquaculture exhibits a broad range of species, and the ability to farm a wide variety of aquatic species reflects the adaptability and versatility of aquaculture practices, allowing for the cultivation of different types of finfish, mollusks, crustaceans, plants, amphibians, and reptiles.

2.1 Aquatic Plants: Farmed aquatic plants include microalgae, also known as phytoplankton, microphytes, and planktonic algae. Macroalgae are commonly known as seaweeds or kelps. Most cultivated taxa are Eucheuma spp., Kappaphycus alvarezii, Gracilaria spp., and Sargassum fusiforme etc. Gracilaria in aquaculture: It is farmed for the primary purpose of producing agar, a valuable substance in various industries. In contrast, other aquatic items mentioned are grown or harvested for direct human consumption and undergo limited processing before being consumed (Sanchez *et al.*, 2018).

Fish farming is a prevalent and important form of aquaculture that involves commercial cultivation of fish in different environments for the primary purpose of food production. It plays a crucial role in meeting the demand for seafood, while contributing to the global economy and food security. The important farmed fish species are carp, salmon, tilapia, and catfish (Troell *et al.*, 2017).

2.2 Crustaceans: Crustacean fisheries are highly diverse in terms of taxa, the majority bof crustacean catch are shrimp. Other than shrimp, lobsters and crabs were also harvested. This industry has played a crucial role in meeting the demand for shrimp in international markets; however, its rapid expansion has also led to environmental and social challenges that have been the focus of ongoing discussions in the aquaculture sector (Gardne *et al.*, 2020).

2.3 Molluscs: Various oyster, mussel, and clam species are examples of aquaculture shellfish, categorized as filter and/or deposit feeders. The unique characteristics of shellfish aquaculture emphasize the filter and deposit feeding behaviors of oysters, mussels, and clams. The generally positive perception of shellfish aquaculture is linked to its potential environmental benefits and low impact on the surrounding ecosystems, making it a sustainable and environmentally friendly form of aquaculture. (Burkholder and Shumway, 2011).

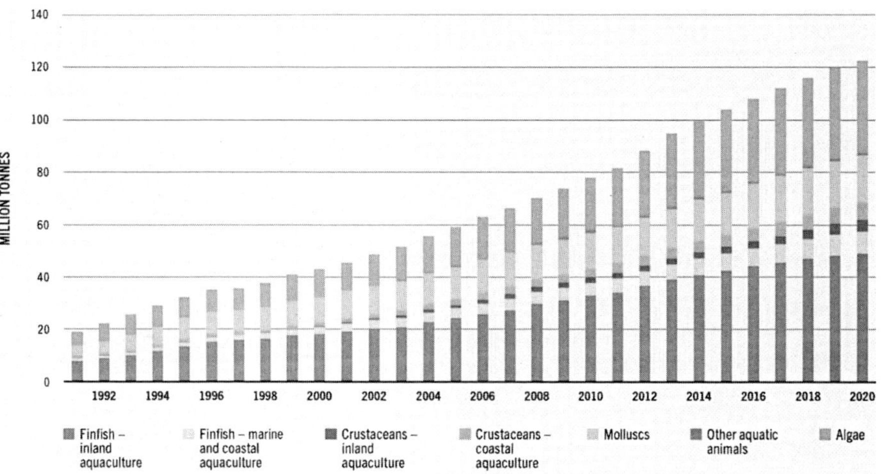

Fig. 4. Global aquaculture production of major species groups and other aquatic animals in million tonnes, 1992–2020, as reported by the FAO. (https://www.fao.org/3/cc0461en/online/sofia/2022/aquaculture-production.html)

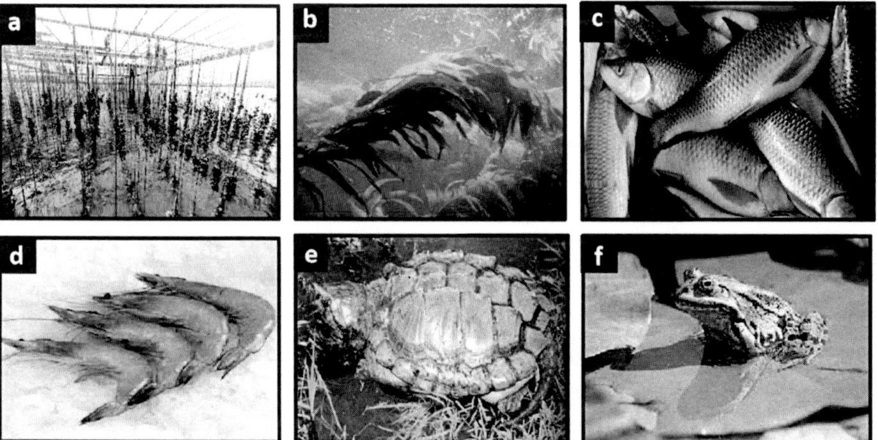

Fig. 5. Image showing a Oyster farming (https://www. istockphoto. com/photos/oyster farming), b Seaweed(https://phys.org/news/2017-09-seaweed-fueled-cars-day-tech.html). Fish (https://www.telegraphindia.com/india/bihar-govt-to-seek-gi-tag-for-mithilas-rohu-fish/cid/1859410) Shrimp (https://7esl.com/crustaceans/e. Turtle (https://nas.er.usgs.gov/taxgroup/reptiles/). Frog (https://aquariussystems.blog/)

Other Groups: Other include aquatic reptiles, amphibians, and miscellaneous invertebrates, such as echinoderms and jellyfish. Inclusion of these groups broadens the scope of aquaculture beyond finfish and shellfish, illustrating the diversity of aquatic species that can be cultivated for different purposes (Iversen & Hale, 2012)

Aquaculture can support the restoration of degraded aquatic ecosystems by providing a source of nutrient-rich water. Aquaculture has several benefits, including providing a sustainable source of seafood, reducing the pressure on wild fish populations, and creating jobs and economic opportunities for local communities. Ongoing research and technological development continue to improve the efficiency and sustainability of various aquaculture systems (Troell *et al.*, 2017).

3. When and Where Aquaculture Started

Fig. 6. Capturing fish in a pond. Engraving from 1582 by Han Bol. (Rijksmuseum, Amsterdam)

Thousands of years after the Neolithic, the origins of aquaculture, particularly fish farming and the cultivation of aquatic plants, can be traced back to ancient China. The practice of raising fish and cultivating aquatic organisms in a systematic manner is believed to have started before 1000 BC. Ancient oracle bones (pieces of turtle shell or bone) as evidence of aquaculture in China mark the origins of aquaculture in a historical and cultural context. Some of the oracle bone inscriptions suggest that the Chinese were involved in systematic sowing and harvesting of fish. These practices have laid the groundwork for more organized and controlled fish farming methods. Over time, aquaculture has become an integral part of Chinese agriculture, contributing significantly to the food supply (Nash *et al.*, 2011). Chinese people are known to cultivate various species of fish, including carp, ponds, and other water bodies. Cultivation of fish such as common carp (Cyprinus carpio) has deep roots in China and could date back to as early as 2000-1000 BC. The long history of aquaculture in China demonstrates the early recognition and cultivation of aquatic resources (Lucas *et al.*, 2019).

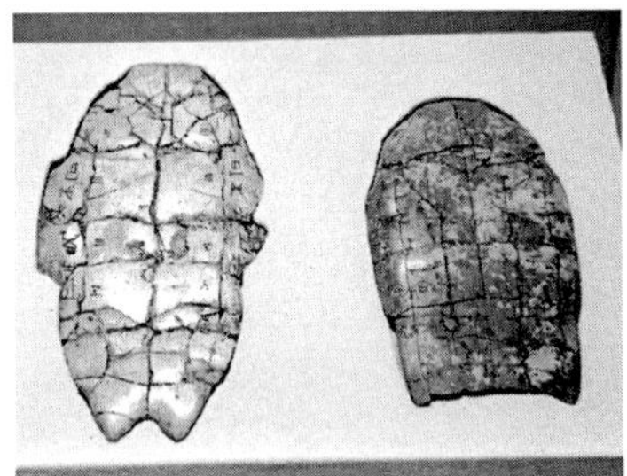

The earliest evidence of aquaculture in China can be found in the marks on ancient "oracle bones." Pieces of turtle shell or bone used in ancient China for divination practices. These inscriptions on oracle bones provide insights into various aspects of ancient Chinese society, including early forms of aquaculture.(Nash *et al.* 2010)

Image a : Oracle bones on display at the National Museum of China in Beijing, March 10, 2023.[Photo/CGTN]

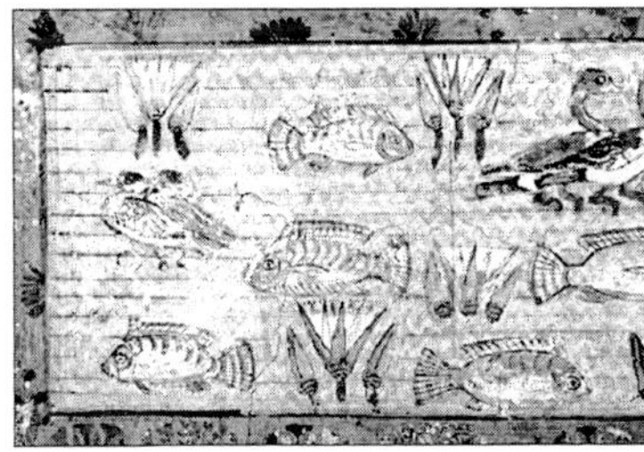

Image b : Central Garden Pool in the Garden of Nebamun's Tomb Painting, British Museum, late 18th Dynasty, circa 1350 BCE (Photo in public domain)- ELECTRUM MAGAZINE

The specific animals such as crocodile, Ibis, Dog-Headed Ape, and Fish are mentioned as being venerated in ancient Egypt.The Egyptians began to keep fishes in sanctuaries and worship them as deities incarnate. This interpretation is based on historical and archaeological studies of ancient Egyptian beliefs and practices.(Wallis *Budge et al.* 1904)

Fig. 7. Showing the facts which indicates the religious importance of aquaculture in ancient time

Evidence suggests that goldfish have been domesticated in China. The goldfish (Carassius auratus) is indeed a close relative of the common carp (Cyprinus carpio) domestication, believed to have started over a thousand years ago during the Tang dynasty (618-907 AD). Historical records indicate that around 970 AD during the Song Dynasty in China, "ponds of mercy" were established to cultivate and raise sacred goldfish. The Chinese regarded them as symbols of good luck, prosperity, and abundance. The establishment of "ponds of mercy' was often associated with Buddhist monasteries, where goldfish were raised and cared for (Kültz *et al.*, 2022).

A recent discovery indicating that the Oujiang color carp, a color variant of the common carp (Cyprinus carpio) known in China for almost 200 years, is genetically closely related or possibly ancestral to Japanese koi (Mabuchi and Song, 2014) Ornamental fish, including goldfish and koi, continue to be popular features in ponds, especially in garden landscapes surrounding temples, prestigious estates, and even private residences in various Asian countries, including Japan, China, and Southeast Asia (Kültz *et al.*, 2022)

Fig. 8. Images showing some examples of koi that display variable patterns of scale color.

Ancient Chinese aquaculture practices have had a lasting impact, not only in China but also globally. China's continued leadership in modern aquaculture has reflected a long history of refinement and expansion of techniques over thousands of years. Aquaculture has become a crucial component of global food production and its roots in ancient Chinese society have played a pivotal role in shaping this industry.

The roots of aquaculture in Northern Africa (specifically Egypt) and Western Asia (the Middle East) may have shared historical connections, particularly during the time of Assyrian rule. The construction of fishponds around sacred places suggests that aquaculture has cultural and possibly religious significance in the development of these ancient societies (Kültz *et al.*, 2022). Assyrian fishponds appear in written documents dating back to 422 bc (Nash *et al.* 2011).

Archaeological evidence of Native American fishing practices provides historical baseline data by defining the biogeography, size, and relative abundance of fish (Gobalet *et al.*, 2012; Lyman *et al.*, 1996; Maschner *et al.*, 2008; Wolverton and Lyman, *et al.*, 2012; Wolverton *et al.*, 2016).

4. Theories on evolution of aquaculture

Aquaculture evolved via a gradual transition from fish hunting to farming. The management of fish by entrapping them in suitable areas and the environment for future use has gradually led to the development of various sophisticated aquaculture systems. Four theories—oxbow, catch-and-hold, concentration, and trap-and-crop—explain how the transition from hunting to farming may have occurred (Rabanal, 1988).

4.1 The oxbow theory proposes that certain oxbows created by rivers mean-dering through the terrain occasionally become isolated from the rivers either during floods followed by droughts or due to shifts in the river course. Capturing more fish than immediately required and storing them in these natural enclosures allowed them to grow larger until needed. Over time, individuals with an entrepreneurial spirit may have strengthened the embankments surrounding oxbow areas and initiated the management of fish in isolated oxbows by introducing fry.

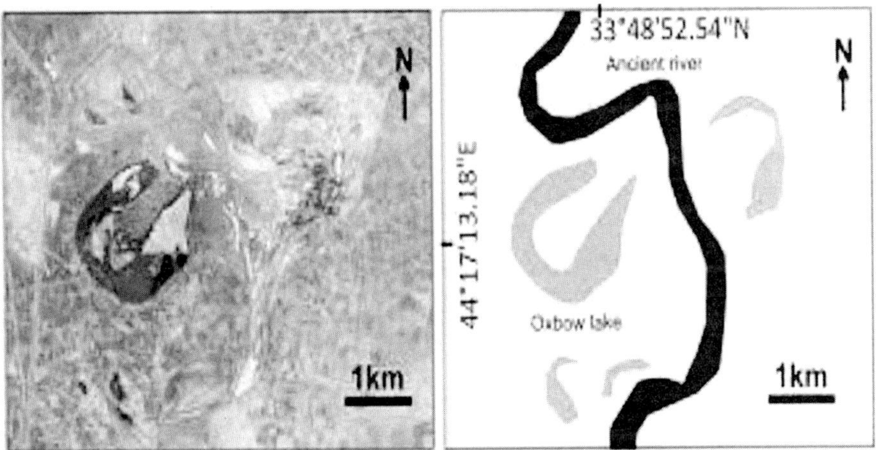

Fig. 9. Image showing oxbow theory of aquaculture. Sources-https://www.researchgate.net/publication/329718098_Recognition_criteria_for_canals_and_rivers_in_the_Mesopotamian_floodplain/figures?lo=1

4.2 The catch-and-hold theory suggests that ponds and moats, initially serving purposes unrelated to aquaculture, such as defending medieval castles in Europe or functioning as mill reservoirs, were subsequently repurposed for multifunctional use-specifically, to detain and cultivate fish as a secondary function. It is conceivable that during peak hunting seasons, when fish supply is abundant, hunters may have disposed of surplus fish in these artificial bodies of water. This practice proved advantageous as certain species thrived in these man-made habitats, making them available for harvest during periods of reduced hunting yields.

This theory is relevant to the dual utilization of dammed storage reservoirs for aquaculture, supplementing their primary role in regulating water flow for operating flour mills in medieval European monasteries (Nash, 2011). Similarly, the simultaneous utilization of floodplains in rice fields may have triggered the inception of aquaculture in Asia (Beveridge and Little, 2014).

Fig. 10. Image showing pond made to defend Castle Sources-https://www.shutterstock.com/image-photo/caerlaverock-castle-moated-triangular-first-built-2252331277

4.3 Concentration theory suggests that extensive plains experiencing natural seasonal flooding, followed by the retreat of floodwater, resulted in the formation of water holes of various sizes within depressions on these plains. As the dry season advanced, these water holes diminished in size, consequently concentrating on the fish they contained. Inhabitants of such floodplains likely seized the opportunity presented by this phenomenon as it simplified the process of catching fish. Over time, it is believed that this practice evolved into the intentional alteration of water holes, including the construction of artificial ponds accompanied by the introduction of fry and supplementary feeding.

4.4 The trap-and-crop theory suggests utilizing brackish and marine habitats for aquaculture purposes. People inhabiting coastal areas slowly learned that several aquatic species regularly enter lagoons, tide pools, and coastal ponds for different purposes. Catching these species in shallow waters is much easier than in the open seas. This knowledge gave rise to the concept of trapping and rearing these aquatic species in coastal ponds by installing gates to prevent the escape of invertebrates and fish that entered and allowed the movement of water.

5. History of aquaculture in India

Historical writings such as Kautilya's Arthashastra (321–300 B.C.) and King Someswara's Manasollasa (1127 A.D.) make references to fish culture. Over centuries, India has upheld the tradition of fish culture in small ponds. Noteworthy advancements in productivity emerged in the early nineteenth century through controlled breeding of carp in tanks designed to simulate river conditions. Brackishwater farming, an age-old practice, involves utilizing man-made barriers in coastal wetlands and salt-resistant deep-water paddy fields. Additionally, in central and southern India, traditional fishing methods have been used for the past 2000 years. Around 300 B.C., Kautilya, in his

"Artha Shastra," provided insights into the potential toxicity of fish in tanks during wartime in India. King Someswara, the son of King Vikramaditya VI, achieved a significant milestone by documenting the common sport fishes of India and categorizing them into marine and freshwater forms in his book Manasoltara, compiled in 1127 AD. In India, around 300 BC, philosopher Kautilya provided insights into the management of fish in freshwater reservoirs (Nash, 2011). Written records dating back to at least 2000 years document the practice of polyculture involving fish, rice, and some other plant species. These ancient origins of aquaculture in India are predominantly centered on the cultivation of freshwater fish, with a primary emphasis on various carp species. In the era of British rule in India, efforts were made to establish sport fisheries by introducing trout into the hill streams of Nilgiris, Kashmir, and the Kulu Valley.

The establishment of fisheries departments plays a pivotal role in promoting the cultivation of both fish and fish. The initial scientifically designed fish farm was built in Sunkesula, Krishna District (now Andhra Pradesh), in 1911 under the supervision of the Madras Fisheries Department. Between 1908 and 1947, Fisheries Departments were established to foster aquaculture development in West Bengal, Punjab, Uttar Pradesh, Andhra Pradesh, and Karnataka.

In the past, fry were sourced from natural waters for aquaculture purposes. The growing demand for seeds to support the expanding aquaculture industry has led to technological advancements in the induced spawning of cultivable species between 1700 and 1900. A significant breakthrough occurred in 1957, when Indian scientists achieved the first successful induced breeding of Indian major carp through hypophysation, followed by the Chinese successfully employing the same technique for Chinese carp in 1958. Similarly, penaeid shrimp species and giant freshwater prawns essential for aquaculture were also successfully hatched under controlled conditions in hatcheries. The establishment of the Pond Culture Division in Cuttack in 1949 marked a significant milestone in the development of freshwater aquaculture in India. This division operated under the name of the Center of the Central Inland Fisheries Research Institute (CIFRI) in West Bengal. On the other hand, brackishwater farming has been a longstanding practice in India, primarily confined to bheries (manmade impoundments in coastal wetlands) in West Bengal and pokkali (salt-resistant deepwater paddy) fields along the Kerala coast. This traditional method involves minimal intervention, mainly the capture of naturally bred juvenile fish and shrimp seeds.

Recognition of the importance of brackishwater aquaculture gained momentum with the initiation of the All India Coordinated Research Project (AICRP) on 'Brackishwater Fish Farming' by ICAR in 1973. Although the project produced various technologies for fish and shrimp farming, current scientific and commercial culture is predominantly focused on shrimp farming. The initial foray into mariculture in India occurred at the Mandapam center of the CMFRI between 1958 and 1959, featuring the culture of milkfish (Chanos chanos). Diverse technologies have been developed for several sedentary species, including oysters, mussels, and clams, as well as for shrimp and finfish. In 1972, CMFRI initiated a successful pearl culture program and developed technology for pearl production in Indian pearl oysters.

Fishing plays a significant role in India's economy, contributing 1.07% of the country's total GDP. (FAO & UN, 2006). This sector supports the livelihoods of more than 28 million people, particularly those in marginalized and vulnerable communities. India is the third largest global fish producer, accounting for 7.96% of the world's production. Additionally, it holds the second position in fish production through aquaculture following China. The estimated total fish production for the fiscal year 2020-21 is 14.73 million metric tons. According to the National Fisheries Development Board, the Fisheries Industry generates export earnings of Rs 334.41 billion, with centrally sponsored schemes expected to boost exports by Rs 1 lakh crore in FY25. Between 2017 and 2020, 65,000 fishermen received training under these schemes. Freshwater fishing constitutes 55% of total fish production. The Ministry of Fisheries, Animal Husbandry, Dairying reported a remarkable growth in fish production, from 7.52 lakh tons in 1950–51 to 125.90 lakh tons in 2018–19, marking a seventeen-fold increase. India commemorates July 10th annually as National Fish Farmers' Day. Notably, the Koyilandy Harbor in Kerala stands as Asia's largest fishing harbor, boasting the longest breakwater.

India has a vast marine coastline spanning 7,516 km (4,670 mi), along with 3,827 fishing villages and 1,914 traditional fish-landing centers. The country's freshwater reservoirs include 195,210 kilometers (121,300 mi) of rivers and canals, 2.9 million hectares of minor and major reservoirs, 2.4 million hectares of ponds and lakes, and approximately 0.8 million hectares of floodplain wetlands and water bodies. As of 2010, the collective sustainable catch fishing potential of marine and freshwater resources exceeded 4 million metric tons of fish. Moreover, India's water and natural resources present a significant growth opportunity in aquaculture (farm fishing), with the potential to achieve ten-fold growth from the 2010 harvest levels of 3.9 million metric tons. This growth is contingent on the adoption of fishing knowledge, regulatory reforms, and sustainability policies in the country. (Roy, Koushik, 2017).

6. The different cultural systems in Indian practice include the following:

1. Intensive pond culture with supplementary feeding and aeration
2. Composite carp culture
3. Weed-based carp polyculture
4. Integrated fish farming with poultry, pigs, ducks, and horticulture.
5. Pen culture
6. Cage culture
7. Running-water fish culture

6.1 Intensive pond culture with supplementary feeding and aeration

In fishponds that receive substantial manure and feeding, and where a considerable portion of the fish feed is generated within the pond, augmenting the flow rate is not a viable solution for enhancing oxygen saturation. In such cases, artificial aeration becomes necessary, involving the elevation of oxygen levels in either the entire pond or specific sections, to guarantee adequate oxygen supply to the fish without constraining production under a specified management level.

In intensive fish farming, ensuring an adequate oxygen supply plays a crucial role in production efficiency. Simultaneously, maintaining saturation levels above 100 percent proves disadvantageous for the normal physiological processes of fish, compelling them to engage in heightened respiration. Consequently, the aeration of fishponds involves the regulation of the dissolved oxygen content in the pond water rather than merely its augmentation. When planning the aeration of fishponds, consideration can be given to expansive water surface areas as potential zones for air intake. The oversaturation of oxygen in the surface water layer can be mitigated through water mixing, promoting a uniform distribution of oxygen throughout the pond. In shallow ponds, such as fishponds, water mixing has no impact on plankton oxygen production. Strategically mixing the water during daytime can ensure the accumulation of significant oxygen reserves for nighttime, when consumption predominates.

Utilization of pure oxygen in intensive fish culture systems

Intensive fish farming systems are typically established in regions with abundant high-quality flow-through water. In these setups, the oxygen needs of the fish were met by the incoming water. Consequently, the fish-carrying capacity of such systems depends on the rate of water flow. However, increasing the water flow beyond a certain threshold faces technical, economic, and physiological constraints, despite the potential for higher stocking densities based on other system elements. Although direct aeration of water can augment fish load to some extent, there are technical and economic limitations to individually aerating fish tanks.

The introduction of pure oxygen removed the constraints on the fish load within the system. This allows for precise regulation of the oxygen supply, enhancing operational safety. A notable advantage of employing pure oxygen is its high mass transfer rate with water, owing to substantial concentration differences. Nonetheless, considerations must be made regarding the costs of oxygen and the investment in dispensing devices. A well-designed and efficiently operated oxygen supply system can offset the increased initial costs owing to its low energy requirements. Pure oxygen is obtainable in gas or liquid form, with liquid oxygen produced through cryogenic separation, and is commonly used in intensive fish farming systems. Stored in vacuum-insulated tanks, 99 percent pure liquid oxygen was vaporized before application. Recent advancements include the development of "oxygen generators, " which are capable of producing 90 percent pure oxygen by eliminating nitrogen from air through a molecular filter.

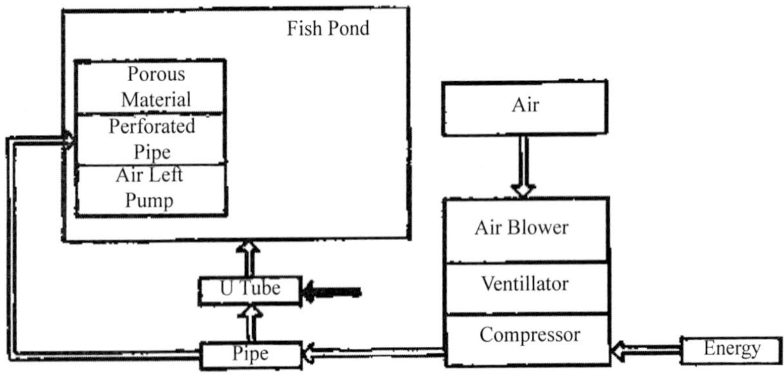

Fig. 11. Image showing the basic principles of fishpond aeration. Air intake type aeration (J.Kepenyes and L. Varadi,Fish Culture Research Institute Szarvas,Hungary)

6.2 Composite carp culture

Composite Fish Culture (https://ccari.icar.gov.in/dss/inlandfish.html) technology involves the simultaneous cultivation of multiple compatible fish species, optimizing pond resources to achieve maximum fish production by utilizing available food organisms in all natural niches. Widely recognized as the most advanced and popular aquaculture technique in the country, it enhances the overall efficiency of fish farming.

Based on the compatibility and feeding habits of the fish, various Indian and Exotic fish species have been identified and recommended for cultivation using Composite Fish Culture technology.

Table 1. Species feeding habits and feeding zone

1. Indian Major Carp

Mrigal	Detritivorus	Bottom feeder
Rohu	Omnivorus	Column feeder
Catla	Zoo plankton feeder	Surface feeder

2. Exotic carps

Grass carp	Herbivorus	Surface, column and marginal areas
Silver carp	Phytoplankton	Surface feeder
Common carp	Detritivorous/omninivorus	Bottom feeder

Fig. 12. Images of the fishes mentioned in Table 1

6.2.1 The key technical considerations for Composite Fish Culture include the following:

6.2.1.1 Pond Selection

- Desilting of existing ponds
- Deepening of shallow ponds
- Excavation of new ponds
- Impoundment of marginal areas of water bodies
- Construction/repairs of embankments
- Construction of inlets/outlets
- Any other necessary items, such as civil structures, watchmen sheds, pump sets, water supply arrangements, and electricity supply arrangements, depend on the project's size and requirements.

6.2.1.2 Pond Management

The management of ponds is crucial for fish farming both before and after the stocking of fish seeds. For new ponds, pre-stocking operations commenced by liming and filling the pond with water. However, for existing ponds in need of development, the initial step involves clearing unwanted weeds and fishes through manual, mechanical, or chemical means. Various methods have been used for this purpose.

i) Manual/Mechanical or Chemical Removal of Weeds: Unwanted weeds are removed either manually, mechanically, or chemically.

ii) Removal of Unwanted and Predatory Fishes and Other Animals: Repeated netting or the use of mahua oil cake 2500 kg/ha or sun drying of the pond bed is employed to remove unwanted and predatory fishes and other animals.

iii) Liming: Ponds with acidic soils/tanks are less productive than are alkaline ponds. Lime was used to adjust pH to the desired level. In addition, lime offers several benefits.

    a) Increase in pH
    b) Acts as a buffer to prevent pH fluctuations.
    c) Enhance soil resistance to parasites.
    d) Exhibit toxic effects, eliminating parasites.
    e) Acceleration of organic decomposition.

6.3 Weed-based carp polyculture

The primary cost of any fish culture operation is attributed to supplementary feeds, leading to an increase in the overall production cost per unit (Mukhopadhyay and Jena, 1999; Fasakin, 1999; New and Csavas, 1993). To mitigate the rise in feed expenses associated with traditional feed items, efforts are underway to explore the utilization of diverse non-conventional feed sources as components of fish feed (Ali *et al.*, 2006). The weed-based system involves incorporating inputs from plant sources, such as weeds, grasses, leaves, or macrophytes, such as duckweeds and Azolla, as supplementary feed in fish production. Herbivorous fish primarily consume these inputs as feed, and subsequently, a portion of the semi-digested fecal matter from the macrophyte-feeding fishes is consumed by other fishes. The remaining portion is recycled within the food chain and serves as a source of nutrients for primary production. Consequently, these nonconventional feed sources have the potential to enhance fish production in aquaculture systems (Grover *et al.* 2000).

6.3.1 Weed based aquaculture system and its history

The weed-based system utilizes inputs from plant sources, such as weeds, grasses, leaves, and macrophytes, such as duckweeds and Azolla, as supplemental feed for fish production. Initially consumed as feed by herbivorous fish, a portion of the semi-digested fecal matter from fish feeding on macrophytes is subsequently consumed by other fish. The remaining portion is recycled within the food chain and serves as a source of nutrients for primary production. This system has the potential to enhance total fish production in aquaculture systems. Aquatic plants have been integral food components in the culture of herbivorous fish for over 4000 years in Egypt and 2500 years in the Orient, including the Indian subcontinent (Bardach *et al.*, 1974). Duckweeds, recognized for their potential as fish food, could play a crucial role in developing cost-effective aquaculture systems in tropical regions (Hassan and Edwards, 1992). Azolla pinnata, traditionally used as a biofertilizer in agriculture, has gained attention in pisciculture because of its ability to fix atmospheric nitrogen and enhance nitrogen levels in semi-intensive pisciculture systems (Ayyappan *et al.*, 1993). Moreover, Azolla serves as a direct food source for certain macrophagous fish (Cassani, 1981). Fresh duckweed (and its dried meal) is suitable for the intensive production of herbivorous fish (Gaiger *et al.*, 1984), and is efficiently converted to live weight gain by carp and tilapia (Van Dyke and Sutton, 1977; Hepher and Pruginin, 1979; Robinette *et al.*, 1980; Hassan and Edwards, 1992; Skillicorn *et al.*, 1993).

Table 2. Aquatic macrophytes used as fish feed in weed based aquaculture

S.No.	Scientific name	Local name	Characteristics
1.	Azolla pinnata	Azolla	The species is typically triangular measuring about 1.5 to 3.0 cm in length, 1 to 2 cm in breadth. Newly form leaves are green but aged leaves are brown in color. With roots.
2.	Spirodela polyrriza	Sonapana	Leaves are flat or oval, 6-10.5 mm in length, 5-10 mm wide and 0.6-1.5 mm thick. Deep green above but with deep brown/reddish ventral. It contains 10-15 roots which are 10-40 mm long.
3.	Lemna minor	Tetulipana	Leaves are flat and elongated, like tamarind tree leaves, 3-4.5 mm in length, 2-2 mm wide and 0.2-0.3 mm thick. Deep green or green in color. It contains single root which is 10-15 mm long.
4.	Wolffia arrhiza	Sujipana, Dimpana	Leaves are minute and rounded, 0.6-1.2 mm in length and 0.5-1 mm wide. Deep green in color and without roots.

Azolla Pinnata

Spirodela Polyrriza

Lemna Minar

Woiffia Arrhiza

Fig. 13. A. Azolla-https://idtools.org/fnwd/index.cfm?packageID=1097&entityID=2558, b. Spirodela by Ulrich Kutschera and Karl J Nikalas, 2014, c. Lemna by Ulrich Kutschera and Karl J Nikalas,2014, d. Wolffia- https://www.knowyourweeds.com/hi/weeds/Wolffia_arrhiza

IIRR-ICLARM (1992) documented successful weed-based integration into aquaculture farming by farmers in various countries including China, Malaysia, Vietnam, the Philippines, and India. In the rural areas of Haryana and Orissa, India, the Indian NGO Sulabh International implemented two duckweed fish production projects (Iqbal, 1999). In 1989, the NGO PRISM-Bangladesh initiated an impressive program aimed at the development and dissemination of duckweed aquaculture in Bangladesh. PRISM has established three centers in Mirzapur, Manikgonj, and Khulna, serving as demonstration farms and training institutions to promote integrated duckweed-fish production in neighboring villages (DWRP, 1996; DWRP, 1997). This historical account underscores the importance of recognizing the endeavors made in the development of web-based aquaculture. This prompts an evaluation of the shortcomings of our research system and necessitates the formulation of strategies for the advancement of cost-effective aquaculture techniques, particularly tailored for resource-poor farmers.

6.4 Integrated fish farming with poultry, pigs, ducks, horticulture, etc.

Integrated fish farming is based on the concept that 'there is no waste, and waste is only a misplaced resource that can become valuable material for another product (FAO, 1977).

Integrated fish farming represents a holistic approach to fish production, in which fish farming is harmoniously combined with other agricultural and livestock operations centered around a fish pond. The various farming subsystems, such as fish, crops, and livestock, are interconnected in a manner that transforms byproducts and wastes from one subsystem into valuable inputs for another. This ensures the comprehensive utilization of land and water resources on farms, leading to maximal and diversified farm output while minimizing financial and labor costs.

Several potential interactions among subsystems exist in a well-designed integrated farming system involving fish, crops, and livestock. For instance, excreta and waste feed from the livestock sub-system serve as both manure and feed for fish, and can also be utilized as crop land fertilizer. Byproducts and wastes from crop cultivation can function as feed and pond manure for fish as well as feed for livestock. Nutrient-rich bottom silt and pond water can serve as effective fertilizers for crop land. This interconnectedness demonstrates that the different subsystems within an integrated framework mutually benefit each other within a confined area, resulting in minimized production costs and a diverse range of outputs, including fish, meat, eggs, vegetables, fruits, fuelwood, and fodder, all essential for the needs of a farm family.

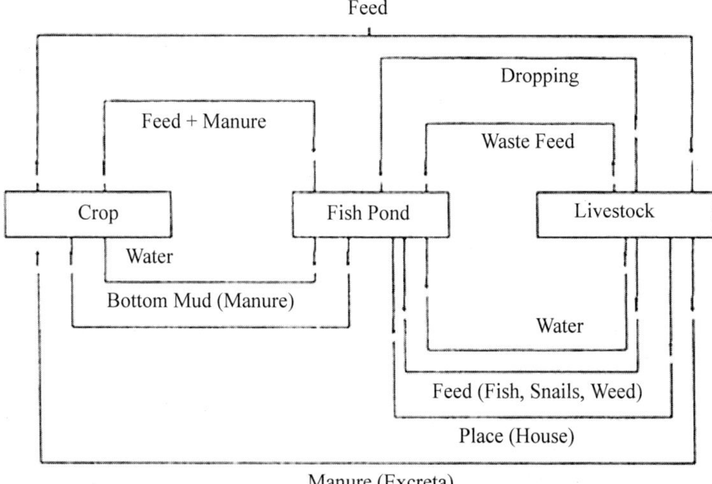

Fig. 14. Interlinking in an integrated fish-crop-livestock farming system (https://www.fao.org/3/ac375e/AC375E04.htm)

6.4.1 Justification of Integrated fish farming for countries like INDIA

Being primarily an agrarian economy, India generates substantial quantities of animal and plant residues, exceeding 322 and 1,000 million metric tons annually, respectively. The country has a significant bovine population, surpassing 307 million cattle heads, along with 181 million sheep and goats, 16 million pigs, and over 150 million poultry and other livestock. In addition to mushroom cultivation, rabbit rearing, sericulture, laculture, and apiculture, which contribute substantial organic material to aquaculture, various agro-based industries also produce effluents that, after appropriate recycling, can be effectively utilized for aquaculture. This is in addition to domestic sewage, which amounts to over 4,000 million liters on a daily basis. (Verma. A.M., 1994)It is crucial to highlight the significant potential for research on integrated farming systems in India, particularly in the Eastern region and specifically in Bihar, across various ecosystems. Despite possessing fertile alluvial soil, a substantial land area under cultivation, abundant sunlight, ample water resources, a considerable livestock population, and a vast human workforce, Eastern India lags significantly behind other regions in Integrated Farming System Research.

In essence, Eastern India represents a high-potential, low-productivity area. There is a substantial opportunity to enhance productivity by leveraging available resources and endowments. Our challenge was to transform adversity into an advantage. The goal is to develop strategies to convert the Indo-Gangetic Basin, focusing predominantly on the lower part, including Eastern Uttar Pradesh, Bihar, and West Bengal, into a major contributor to the country's agricultural output. Integrated farming systems in conjunction with modern technologies can play a pivotal role in achieving this objective. Thus, it is imperative to dedicate attention and resources to fortify the Integrated Farming System Research Programme both nationally and specifically in the eastern part of the country (Kumar et al 2012).

Fig. 15. Image of integrated fish farming (https://www.asiafarming.com/integrated-farming-fish-livestock)

6.5 Pen culture

Pen culture involves the cultivation of fish within a confined volume of water enclosed on all sides except the bottom, allowing for the unrestricted flow of water from at least one side. This method can be viewed as a hybrid approach that combines the elements of both pond and cage cultures. Typically, shallow areas along the shores and banks of lakes and reservoirs are used to construct pens or enclosures using nets or wooden materials, providing a space for fish cultivation.

In a fishpen, the natural lake bottom serves as the floor of the enclosure. One key advantage of pens is their ability to contain benthic fauna, which serve as a food source for fish. Similar to pond culture and polyculture, simultaneous cultivation of multiple species can be implemented in pens. The environment within a fish pen is characterized by the free exchange of water with the surrounding water body, ensuring high concentrations of dissolved oxygen.

Pens are utilized for diverse aquaculture endeavors in various countries, including the Philippines, Indonesia, Thailand, Malaysia, China, and the USA (Beveridge 1984; Chua and Teng 1977; Shang and Tisdell 1997). In India, attempts have been made to rear carp seed through pen culture in oxbow lakes, swampy tanks, beels, and reservoirs (Abraham 1980; Banerjee and Pandey 1978; Swaminathan and Singit 1982; Yadava *et al.,* 1983). However, this technology has not yet been standardized.

Fig. 16. Pen culture (https://aquafind.com/articles/Modified_Aquaculture_Systems_In_India.php)

Natarajan (1976) and Selvaraj *et al.* (1990) proposed the introduction of advanced fingerlings (exceeding 100 mm in length) into reservoirs to enhance survival rates and fish production. Despite this recommendation, development agencies persist in stocking smaller fish seed (15-40 mm) because of the limited rearing space in land-based ponds as the construction of new ponds involves significant capital investment. In such circumstances, the adoption of fish culture in pens is appealing, as these pens can be easily set up by unskilled labor, and the materials needed for their construction are both inexpensive and readily available in local markets.

6.6 Cage culture

Cage culture (Das *et al.*, 2009) involves the cultivation of fish from fry to fingerling, fingerling to table size, or table size to marketable size within a confined space that allows for unrestricted exchange of water with the surrounding water body. The enclosure, commonly referred to as a cage, is completely surrounded by mesh netting crafted from a synthetic material, specifically designed to withstand decomposition in water over an extended period. This material is often marketed under the brand name, Netlon. Typically, these cages are relatively small, ranging from 1 square meter (m2) to 500 m2 in freshwater reservoirs. For a more intensive culture, a combination of several small cages organized in a battery, as detailed below, proved to be suitable.

Fig. 17. Cage culture (five cages and two cages under All India Network Project on Mariculture of ICAR- Central Marine Fisheries Research Institute (CMFRI), Karwar, India.)

India's reservoirs, covering a combined surface area of 3.25 million hectares (ha), predominantly in the tropical zone, stand as the nation's most crucial inland water resource, holding substantial untapped potential. Achieved fish yields included 50 kg/ha/year from small reservoirs, 20 kg/ha/year from medium-sized reservoirs, and 8 kg/ha/year from large reservoirs. These figures still allow room for increasing fish yields through capture fisheries, including culture-based approaches. Auto-stocking success rates are notably low in Indian reservoirs, particularly in smaller reservoirs, where many may partially or completely dry up during the summer, leading to no surviving stock. Implementing a regular, well-planned, and sustained stocking strategy could significantly boost fisheries in these water bodies. The central aim of the cage culture is to rear fingerlings, especially carp, measuring over 100 millimeters (mm) in length for subsequent stocking in reservoirs.

References

Abraham, M. (1980). Rearing of spawn in pen. J. Inland Fish. Soc. India, 12(2), 97-100.

Area Handbook, series (1996). Area Handbook Series, American University (Washington, D.C.). Foreign Area Studies Volume 550 of DA pam. United States: American University, Foreign Area Studies. pp. 416. ISBN 0-8444-0833-6.

Ayyappan, S., Dash, B., Pani, K. C. and Tripathi, S. D. 1993. Azolla – a new aquaculture input, National meet on Aquafarming system practices and potential, Abstract No. OFS-2, 10-11 February, 1993, CIFA, Bhubaneswar, India

Banerji, S. R., & Pandey, K. K. (1978). Carp seed stocking in an ox-bow lake of Champaran (Bihar) through rearing enclosure. J. Inland Fish. Soc. India, 10, 131-134.

Barnabe, G. (2018). Aquaculture: biology and ecology of cultured species. CRC Press.

Bardach, J. E., Ryther, J. H., & McLarney, W. O. (1974). Aquaculture: the farming and husbandry of freshwater and marine organisms. John Wiley & Sons.

Beveridge, M. C. (1984). Cage and pen fish farming: carrying capacity models and environmental impact (No. 255-259). Food & Agriculture Org.

Beveridge, M. C., & Little, D. C. (2002). The history of aquaculture in traditional societies. Ecological aquaculture. The evolution of the Blue Revolution, 3-29.

Budge, E. A. W. (1904). The Gods of the Egyptians: Or, Studies in Egyptian Mythology (Vol. 1). Methuen & Company.

Burkholder, J. M., & Shumway, S. E. (2011). Bivalve shellfish aquaculture and eutrophication. Shellfish aquaculture and the environment, 155-215.

Cassani, J. R. (1981). Feeding behaviour of underyearling hybrids of the grass carp, Ctenopharyngodon idella♀ and the bighead, Hypophthalmicthys nobilis♂ on selected species of aquatic plants. Journal of Fish Biology, 18(2), 127-133.

Chua, T. E. Floating fishpens for rearing fishes in coastal waters, reservoirs and mining pools in Malaysia/by Chua Thia-Eng and Teng Seng Keh.

Das, A. K., Vass, K. K., Shrivastava, N. P., & Katiha, P. K. (2009). Cage culture in reservoirs in India (A Handbook). WorldFish.

"Fisheries". Tamil Nadu Agricultural University, Coimbatore. 2007.

"India - National Fishery Sector Overview".2006 Food and Agriculture Organization of the United Nations.

Gaigher, I. G., Porath, D., & Granoth, G. (1984). Evaluation of duckweed (Lemna gibba) as feed for tilapia (Oreochromis niloticus× O. aureus) in a recirculating unit. Aquaculture, 41(3), 235-244.

Gardner, C., Watson, R. A., Jayanti, A. D., Suadi, J., AlHusaini, M., Lovrich, G., & Thiel, M. (2020). Crustaceans as fisheries resources: general overview. Fisheries and Aquaculture: Volume 9, 9.

Gobalet, K. W. (2012). A native Californian's meal of coho salmon (Oncorhynchus kisutch) has legal consequences for conservation biology. Exploring Methods of Faunal Analysis: Insights from California Archaeology, 87-95.

Hassan, M. S. and Edwards, P. 1992, Evaluation of duck weed (Lemna perpusilla and Spirodela polyrhiza) as feed for Nile Tilapia (Oreochromis niloticus). Aquaculture, 104: 315-326.

Hepher, B., & Pruginin, Y. (1979). Guide to fish culture in Israel. 4. Fertilisation, maturing and feeding. Foreign Training Dept., Israel.

Iqbal, S. (1999). Duckweed aquaculture. Potentials, possibilities and limitations for combined wastewater treatment and animal feed production in developing countries. SAn-DEC Report, (6/99).

Iversen, E. S., & Hale, K. K. (2012). Aquaculture sourcebook: A guide to North American species. Springer Science & Business Media.

Kepenyes, J., & Váradi, L. (1984). Aeration and oxygenation in aquaculture. Inland Aquaculture Engineering, FAO, Rome, Italy, 473-505.

Kültz, D. (2022). A Primer of Ecological Aquaculture. Oxford University Press.

Kumar, S., Dey, A., Kumar, U., Chandra, N., & Bhatt, B. P. (2012). Integrated farming system for improving agricultural productivity. Status of Agricultural Development in Eastern India, 205-230.

Lucas, J. S., Southgate, P. C., & Tucker, C. S. (Eds.). (2019). Aquaculture: Farming aquatic animals and plants. John Wiley & Sons.

Lyman, R. L. (1996). Applied zooarchaeology: The relevance of faunai analysis to wildlife management. World Archaeology, 28(1), 110-125.

Mabuchi, K. and Song, H. (2014). 'The complete mitochondrial genome of the Japanese ornamental koi carp (Cyprinus carpio) and its implication for the history of koi', Mitochondrial DNA, 25, pp. 35–36. https://doi.org/10.3109/19401736.2013.779261

Maschner, H. D., Betts, M. W., Reedy-Maschner, K. L., & Trites, A. W. (2008). A 4500-year time series of Pacific cod (Gadus macrocephalus) size and abundance: archaeology, oceanic regime shifts, and sustainable fisheries.

Md.Asadujjaman(2014)Development of low cost weed based aquaculture technology in ponds RUCL Institutional Repository http://rulrepository.ru.ac.bd

Murugesan, V. K., Manoharan, S., & Palaniswamy, R. (2005). Pen fish culture in reservoirs: An alternative to land based nurseries.

Nash, C. (2010). The history of aquaculture. John Wiley & Sons.

Natarajan, A. V. (1976). Ecology and the state of fishery development in some of the man-made reservoirs in India.

Rabanal, H. R. (1988). History of aquaculture.

Roy, Koushik (2017). "Technicalities to be considered for culture fisheries development in Indian inland waters: seed and feed policy review". Environment, Development and Sustainability. 21: 281–302. doi:10.1007/s10668-017-0037-3. S2CID 158126876.

Robinette, H. R., Brunson, M. W. and Day, E. J. 1980. Use of duck weed in diets of channel catfish. Proc. 13th Ann. Conf. SE Assoc. Fish Wildlife Age, 108-114 pp.

Sanchez, G. M., Gobalet, K. W., Jewett, R., Cuthrell, R. Q., Grone, M., Engel, P. M., & Lightfoot, K. G. (2018). The historical ecology of central California coast fishing: Perspectives from Point Reyes National Seashore. Journal of Archaeological Science, 100, 1-15.

Selvaraj, C., Murugesan, V. K., & Aravindakshan, P. K. (1990). Impact of stocking of advanced fingerlings on the fish yield from Aliyar reservoir. Contributions to the Fisheries of Inland Open Water Systems in India Part I, IFSI, CIFRI, Barrackpore, India, 109-114.

Shang, Y.C. and C.A. Tisdell. 1997.Economic decision making in sustainable aquaculture development,p. 127-148. In J.E. Bardach (ed.)Sustainable Aquaculture John Wiley& Sons, Inc.

Skillicorn, P., Spira, W. and Journey, W. 1993. Duckweed Aquaculture- A New Aquatic Farming System for Developing Countries. The World Bank, Washington DC. 76p.

Swaminathan, V., & Singit, G. S. (1982). Massive fish seed production in pens: major experiment underway at Tungabhadra dam. Fishing Chimes, 2(7), 42-45.

Tacon, A. G. (2020). Trends in global aquaculture and aquafeed production: 2000–2017. Reviews in Fisheries Science & Aquaculture, 28(1), 43-56.

Tacon, A. J. (2003). Aquaculture production trends analysis. Review of the state of world aquaculture. FAO fisheries circular, 886(2), 5-29.

Troell, M., Kautsky, N., Beveridge, M., Henriksson, P., Primavera, J., Rönnbäck, P., ... & Jonell, M. (2017). Provided for non-commercial research and educational use. Not for reproduction, distribution or commercial use.

Van Dyke, J. M., & Sutton, D. L. (1977). Digestion of duckweed (Lemna spp.) by the grass carp (Ctenopharyngodon idella). Journal of Fish Biology, 11(3), 273-278.

Wolverton, S. J., Rick, T. C., & Nagaoka, L. (2016). Applied zooarchaeology: five case studies. Applied Zooarchaeology, 1-130.

Wolverton, S., & Lyman, R. L. (Eds.). (2012). Conservation biology and applied zooarchaeology. University of Arizona Press.

Yadava, Y. S., Choudhury, M., Kolekar, V., Singh, R. K., & Natarajan, P. (1983). Pen Farming-utilizing marginal areas of beels in Assam for carp culture. In Proceedings of the National Seminar on Cage Pen Culture, Fisheries College and Research Institute, Tuticorin, India (pp. 55-58).

2

The Current Status of Aquaculture and Fisheries: Challenges and Opportunities

Felegush Erarto[1*], Durgesh Kumar Verma[2] and Sanjay Kumar Gupta[3]

[1]Department of Fisheries and Aquatic Sciences, Bahir Dar University, Ethiopia

[2]ICAR-Central Inland Fisheries Research Institute, Regional Centre Prayagraj-211002, Uttar Pradesh, India

[3]ICAR-Indian Institute of Agricultural Biotechnology, Namkum, Ranchi-834010, India

Abstract

Fisheries are capture fisheries, which involve catching wild fish from natural bodies of water, and culture fisheries, also known as aquaculture, include raising fish in controlled environments. Aquaculture offers a promising avenue for meeting the growing demand for seafood. Besides, capturing fisheries has been a traditional seafood source for centuries. Some of the aquaculture's advantages are increased food production (i.e., producing over half of all seafood consumed globally) and improved food security, particularly in developing nations where protein deficiency and malnutrition pose significant challenges. While fisheries face overfishing, habitat destruction, and by-catch, aquaculture contends with environmental impacts, disease management, and feed sustainability. Nevertheless, both sectors present promising solutions: fisheries through sustainable management, technological advancements, marine protected areas, aquaculture through increased food production, reduced environmental impact, economic benefits, and sustainable feed sources. By addressing these challenges and embracing these opportunities, fisheries and aquaculture can contribute to a more sustainable and food-secure future. Therefore, the future of seafood production will likely involve a combination of aquaculture and capture fisheries, each playing a complementary role by addressing their respective challenges and embracing opportunities for improvement. Aquaculture and capture fisheries can work together to ensure a sustainable seafood supply for a growing global population.

Keywords: *aquaculture, capture, fisheries, food security, opportunities*

1. Introduction

The global population is predicted to increase from 6.8 billion (now) to 9 billion by 2050 (Garcia and Rosenberg 2010). It intensified competition for finite resources (e.g., land and water), exacerbating food insecurity and increasing global food demand (Ochieng and Munguti 2016). Ensuring food security is a paramount challenge of our era (Carrera-Quintana *et al.*, 2022), though it is critical for the world's sustainable development. The world's rising population and expanding middle class are driving up demand for nutritious, healthy, and protein-rich foods (FAO, 2018; Duggan and Kochen 2016). The least expensive source of high-quality animal protein is fish (Hasselberg *et al.*, 2020), vitamins and omega-3 fatty acids, etc. (Nelson *et al.*,, 2016; Ghose 2014). As a result, fish and fishery products play a significant role in international trade; in 2012, the total export production rose from 25% in the middle of the 1970s to around 37% (Bellmann *et al.*, 2016).

Fisheries have two categorizations: finfish fisheries and non-finfish fisheries. Finfish fisheries focus on harvesting and raising actual fish species, e.g., salmon, trout, etc. These fisheries can be divided into two types: capture fisheries, which involve catching wild fish from natural water bodies and culture fisheries, also known as aquaculture, which include raising fish in controlled environments, such as ponds, tanks, and cages. Non-fin fisheries target other aquatic organisms, including prawns, crabs, lobsters, mussels, oysters, sea cucumbers, frogs, and seaweeds. Both large-scale fisheries and aquaculture play significant roles in global development (Finegold 2009). Millions of people rely on them for food and a living, with inland capture fisheries and aquaculture accounting for more than 40% of the world's reported finfish production. These fishing industries are crucial to reducing poverty and ensuring food security (Lynch *et al.*, 2016).

While fisheries and aquaculture are the most cost-effective ways to obtain high-quality protein and vital minerals such as vitamins, omega-3 fatty acids, and amino acids, they deserve more attention in national development strategies and donor funding priorities. For the purpose of protecting capture fisheries and guaranteeing that aquaculture is sustainable and advantageous to low-income individuals, appropriate laws and regulations are still essential (Finegold 2009). For instance, the production of inland water capture reached 11.6 million tonnes globally in 2012. Although it seems to be continuously growing, its proportion of worldwide capture production is about 13 percent. The most challenging subsector is obtaining dependable statistics on catch productivity in inland waters (Bediye 2022). The catch is declining quickly for several reasons, including illegal fishing, overfishing, population growth, pollution, the introduction of exotic species, climate change, land degradation, and other environmental factors. Even though capturing fisheries is the most essential contributing sector for developed and developing countries, the stock still needs to grow.

2. Opportunities of capture fishery

Wild-caught or capture fisheries are vital to the world's food and economic prosperity. It has the enormous potential to benefit future generations economically and ensure sustained food security. Nevertheless, several issues this industry faces jeopardize its viability and limit its ability to improve human welfare on a national level (Laghari 2018). The leading causes of native fish reduction include investment, lack of understanding of fish ecology, and water quality. Human activity, anthropogenic disturbance, and ignorance are the three leading causes of fish extinction globally (Gurung 2013). Generally, the four main human activities, habitat destruction, introduction, pollution, and overexploitation of resources, also collectively called HIPO, negatively impact fish and fisheries. Fish population decreases are frequently attributed to habitat alteration. The bottom topography and above-bottom structure are essential for the survival of fish species, and these factors can be changed through channelization, dam construction, disturbance of watersheds, competition for water, and other means.

The second threat of capturing fisheries is introducing exotic species intentionally to enhance fisheries (e.g., common carp in Lake Victoria), biological control, recreational use, aquaculture, or unintentional. Introducing alien species poses a second concern to catch fisheries; species dispersal is a normal zoogeographic process. It can happen accidentally or unintentionally to improve aquaculture, biological control, recreational use, or fisheries (like the common carp in Lake Victoria). Native species may experience population declines or extinction due to introductions; this can happen directly through predation on adults, eggs, and young or indirectly through pathogen transfer, competition, or hybridization (Erarto and Getahun 2020).

The third threat to various inland waters and their fisheries is pollution. It can come in sediments, dissolved or suspended materials, runoff, or precipitation. Pollution includes elemental contaminants like heavy metals, chlorine, and chemical complexes. As a result, direct toxicity, or food chain consequences, harms fish health, ultimately compromising fish survival and reproduction. The pollutants originate from either point sources, easily identified and linked to a specific area, or non-point sources, which are more challenging to identify and link to specific location sources (Gebremariam *et al.*, 2002). Pollutants also lead to changes in water's physical, chemical, and biological parameters, which distress fisheries and the functioning of the whole aquatic ecosystem. Overfishing is another fourth and most pressing challenge facing capture fisheries. It occurs when fishing activity exceeds the reproductive capacity of fish populations, leading to stock depletion and potential collapse. Overfishing is driven by various factors, including increasing demand for seafood, technological advancements in fishing gear, and subsidies that incentivize excessive fishing efforts (Garcia and Newton 1995; Chuenpagdee *et al.*, 2005; Pramulati *et al.*, 2023). Illegal fishing is a widespread, global phenomenon affecting already heavily depleted wild fish stocks and threatening habitats. It contributes to water quality deterioration through discarded fishing gear (Vince *et al.*, 2020).

Generally, addressing challenges and seizing opportunities requires concerted efforts from governments, international organizations, the fishing industry, and consumers. By adopting sustainable practices, promoting responsible governance, and practicing conservation, it is possible to ensure that capture fisheries continue to nourish the world while preserving the health of aquatic ecosystems. In addition, using aquaculture as an alternative means of preventing overfishing in capture fisheries and shifting some of the demand from wild-caught seafood to farmed seafood will reduce the pressure on wild fish stocks and promote sustainable fisheries management.

2.1 Aquaculture as an alternative means for food security

As capture fisheries production is declining for several reasons, the interest in aquaculture species is increasing to fulfill the aquatic protein demand nationwide. Aquaculture offers several advantages: sustainable food production, employment and income opportunities, resource efficiency, diet diversification through various seafood options, promotion of research and innovation, reduced by-catch, conservation ecosystem, and food security. Most importantly, it is commonly known as an alternative means of meeting seafood demand, which will continue to grow with population growth, rising incomes, and urbanization (Finegold 2009). Globally, aquaculture provides over 50% of fish for human consumption and is expected to continue to increase long-term (Niyibizi *et al.*, 2022). Its annual growth rate is about 6%, and its contribution is more significant than that of wild fisheries (FAO, 2018). Furthermore, aquaculture's contribution is expected to increase further (Shen *et al.*, 2021). Aquaculture, an efficient protein food production system from the aquatic environment, has been practiced worldwide (Bediye 2022). Chinese fisheries production embarked on a remarkable growth, commencing with 4.3 million metric tons in 1978, reaching a milestone of 12.4 million metric tons in 1990, and culminating in an impressive 47.48 million metric tons in 2008 and also; a similar incremental was observed in lake fisheries over the same period (Jia *et al.*, 2013). Aquaculture production in Brazil also in 2019 was estimated at 800,000 tonnes, representing a gross revenue of US$ 1 billion (Valenti *et al.*, 2021)

Aquaculture has been recognized as an essential opportunity to enhance household food security in developing nations (Ochieng and Munguti 2016). Additionally, it contributes to international fish trade, global food security, and poverty alleviation because catching wild fish from lakes, dams, and oceans cannot satisfy the country's present need for wholesome food (Ababouch *et al.*, 2023). Approximately 6.2 million people are employed in aquaculture in Africa, and a significant percentage of these workers are women engaged in large-scale commercial operations (Satia, 2016). Women primarily engage in the aquaculture value chain's downstream postharvest and marketing operations. Therefore, the aquaculture industry has the potential to significantly advance African economic growth, lower unemployment rates, and increase food security (Adeleke *et al.*, 2020). Egypt, Nigeria, Uganda, Ghana, Tunisia, Kenya, Zambia, Madagascar, Malawi, and South Africa are the leading aquaculture producers in Africa (Satia 2016). Over the past ten years, these major aquaculture producers have grown significantly as a result of several factors, including the promotion of private sector-led aquaculture development, capacity building in critical subject areas, adoption of good governance, research and development, and credit facility accessibility (Satia, 2017). Nowadays, the Ethiopian government has also suggested aquaculture to increase fish production and as a substitute for ensuring food security and reducing poverty. It is currently considered an essential component of plans and strategies for rural and agricultural development (Chimdo 2022).

2.2 Challenges of aquaculture

Despite the numerous advantages of aquaculture, it also comes with several challenges that must be carefully considered and addressed to ensure its sustainability and environmental responsibility. These challenges include water pollution, habitat loss, species introduction, antibiotic overuse, dependency on non-sustainable feed sources, community uprooting, and genetic homogeneity within farmed fish populations are some of these issues. Globally, aquaculture is susceptible to the spread of diseases among farmed fish due to the prevalence of bacterial, viral, and other infectious diseases. While these diseases are being detected and concurrently treated with a range of therapeutic and prophylactic measures, the broad-spectrum efficacy of vaccines stands out as a crucial preventative approach in aquaculture. However, the effectiveness of treatments such as antibiotics and probiotics appears to diminish as novel mutant strains emerge and disease-causing pathogens resist conventional antibiotics (Mondal and Thomas 2022). Furthermore, the overuse of antibiotics can result in the emergence of germs that are resistant to bacteria. This is a serious concern, as it can make treating diseases in humans and animals difficult.

Another challenge of aquaculture is feed availability, and aquaculture relies on fishmeal and fish oil, derived from wild-caught fish, as primary feed ingredients (Rana 2013). The availability of suitable artificial feeds made using inexpensive, locally sourced ingredients that can meet the nutritional needs of the fish being cultivated is a vital component of fish farming's success. Because of its high protein content, balanced amino acid profile, mineral and vitamin content, palatability, and digestibility, fishmeal (FM) is frequently used as the preferred protein source in aqua-feeds. However, the overreliance on FM as the primary protein source in fish feeds has increased scarcity, costliness, and unsustainability (Musyoka *et al.*, 2020). Aquaculture also has a significant impact on the surrounding environment. One of the primary concerns is water pollution. Fish farms release excess nutrients, such as nitrogen and phosphorus, into the surrounding water bodies. These excess nutrients can lead to eutrophication, a condition characterized by excessive plant growth that can deplete oxygen levels and harm aquatic life. Algal blooms, which are dense concentrations of algae, can also occur due to eutrophication. These blooms can further reduce oxygen levels and release harmful toxins to humans and animals. Intensive aquaculture practices can lead to reduced genetic diversity among farmed fish populations. This can make them more susceptible to diseases and reduce their resilience to environmental changes. If a disease were to spread through a genetically homogenous population, it could wipe out the entire population. Generally, aquaculture's challenges are complex and interconnected, so a holistic approach is needed. This approach should include effective regulations, environmental monitoring, technological advancements, and responsible aquaculture practices. Implementing responsible aquaculture practices that minimize environmental impacts, promote disease prevention, utilize sustainable feed sources, and address social concerns are essential for the long-term sustainable and environmentally responsible source of food production. According to Mzula *et al.* (2021), for sustainable aquaculture to be practiced, fish producers must have access to extension services and well-established regulatory frameworks for optimal pond management practices and fish health.

Summary

Aquaculture and capture fisheries both have their own sets of challenges and opportunities. Aquaculture offers promise in meeting the growing demand for seafood but also negatively impacts the environment and food sustainability. Capture fisheries provide traditional seafood sources and economic benefits but face challenges like overfishing, habitat degradation, and pollution. The future of seafood production lies in a combination of aquaculture and capture fisheries, where each sector addresses its challenges and improves its practices.

References

Ababouch, L., K. A. T. Nguyen, M. Castro de Souza and J. Fernandez-Polanco (2023). Value chains and market access for aquaculture products. Journal of the World Aquaculture Society 54(2): 527-553.

Adeleke, B., D. Robertson-Andersson, G. Moodley and S. Taylor (2020). Aquaculture in Africa: A Comparative Review of Egypt, Nigeria, and Uganda Vis-À-Vis South Africa. Reviews in Fisheries Science Aquaculture 29(2): 167-197.

Bediye, D. T. (2022). Challenges and Prospects of Capturing Fisheries and the Ways Forwards in Developing Countries. J Fisheries Livest Prod, 10, 355..

Bellmann, C., A. Tipping and U. R. Sumaila (2016). Global trade in fish and fishery products: An overview. Marine Policy 69: 181-188.

Boyd, C. E., L. R. D'Abramo, B. D. Glencross, D. C. Huyben, L. M. Juarez, G. S. Lockwood, A. A. McNevin, A. G. J. Tacon, F. Teletchea, J. R. Tomasso, C. S. Tucker and W. C. Valenti (2020). Achieving sustainable aquaculture: Historical and current perspectives and future needs and challenges. Journal of the World Aquaculture Society 51(3): 578-633.

Carrera-Quintana, S. C., P. Gentile and J. Girón-Hernández (2022). An overview on the aquaculture development in Colombia: Current status, opportunities and challenges. Aquaculture 561.

Chamberlain, G., and Rosenthal, H. (1995). Aquaculture in the next century: opportunities for growth, challenges of sustainability. World Aquaculture, 26(1), 21-25..

Chan, C. Y., N. Tran, K. C. Cheong, T. B. Sulser, P. J. Cohen, K. Wiebe and A. M. Nasr-Allah (2021). The future of fish in Africa: Employment and investment opportunities. PLoS One 16(12): e0261615.

Chimdo, A. (2022). Review on potential and challenges of aquaculture practice in Ethiopia. Applied Water Science 12(9).

Chuenpagdee, R., Degnbol, P., Bavinck, M., Jentoft, S., Johnson, D., Pullin, R., and Williams, S. (2005). Challenges and concerns in capture fisheries and aquaculture. Fish for life: interactive governance for fisheries. Amsterdam University Press, Amsterdam, The Netherlands, 25-40.

Chuenpagdee, R., Kooiman, J., and Pullin, R. (2008). Assessing governability in capture fisheries, aquaculture and coastal zones. Journal of Transdisciplinary Environmental Studies, 7(1), 1-20..

Duggan, D. E. and M. Kochen (2016). Small in scale but big in potential: Opportunities and challenges for fisheries certification of Indonesian small-scale tuna fisheries. Marine Policy 67: 30-39.

Erarto, F., and Getahun, A. (2020). Impacts of introductions of alien species with emphasis on fishes. Int J Fish Aquat Stud., 8(5), 207-216.

FAO. 2018. The State of World Fisheries and Aquaculture 2018 –Meeting the Sustainable Development Goals. Rome. 224 pp.

Finegold, C. (2009). The importance of fisheries and aquaculture to development. The Royal Swedish Academy of Agriculture and Forestry..

Garcia, S. M., and Newton, C. (1995). Current situation, trends and prospects in world capture fisheries. FAO, Fisheries Department.

Garcia, S. M. and A. A. Rosenberg (2010). Food security and marine capture fisheries: characteristics, trends, drivers and future perspectives. Philos Trans R Soc Lond B Biol Sci 365(1554): 2869-2880.

Gebre-Mariam, Z., and Desta, Z. (2002). The chemical composition of the effluent from Awassa textile factory and its effects on aquatic biota. SINET: Ethiopian Journal of Science, 25(2), 263-274.

Ghose (2014). Fisheries and aquaculture in Bangladesh: Challenges and opportunities. Annals of Aquaculture and Research, 1(1), 1-5..

Grafton, R. Q., Hilborn, R., Squires, D., and Williams, M. J. (2010). Marine conservation and fisheries management: at the crossroads (Vol. 1). Chapter.

Gurung, T. B. (2013). Native fish conservation in Nepal: Challenges and opportunities. Nepalese Journal of Biosciences 2: 71-79.

Hasselberg, A. E., I. Aakre, J. Scholtens, R. Overå, J. Kolding, M. S. Bank, A. Atter and M. Kjellevold (2020). Fish for food and nutrition security in Ghana: Challenges and opportunities. Global Food Security 26.

Jia, P., W. Zhang and Q. Liu (2013). Lake fisheries in China: Challenges and opportunities. Fisheries Research 140: 66-72.

Laghari, M. Y. (2018). Aquaculture in Pakistan: Challenges and opportunities. International Journal of Fisheries and Aquatic Studies, 6(2), 56-59..

Lakew, A., Dagne, A., Tadesse, Z. (2018). Fishery and aquaculture Research in Ethiopia: Challenges and future directions. Agricultural Research for Ethiopian Renaissance.

Lynch, A. J., S. J. Cooke, A. M. Deines, S. D. Bower, D. B. Bunnell, I. G. Cowx, V. M. Nguyen, J. Nohner, K. Phouthavong, B. Riley, M. W. Rogers, W. W. Taylor, W. Woelmer, S.-J. Youn and T. D. Beard (2016). The social, economic, and environmental importance of inland fish and fisheries. Environmental Reviews 24(2): 115-121.

Lynch, A. J., V. Elliott, S. C. Phang, J. E. Claussen, I. Harrison, K. J. Murchie, E. A. Steel and G. L. Stokes (2020). Inland fish and fisheries integral to achieving the Sustainable Development Goals. Nature Sustainability 3(8): 579-587.

Mondal, H. and J. Thomas (2022). A review on the recent advances and application of vaccines against fish pathogens in aquaculture. Aquac Int 30(4): 1971-2000.

Munguti, J. M., J.-D. Kim and E. O. Ogello (2014). An Overview of Kenyan Aquaculture: Current Status, Challenges, and Opportunities for Future Development. Fisheries and aquatic sciences 17(1): 1-11.

Mzula, A., P. N. Wambura, R. H. Mdegela and G. M. Shirima (2021). Present status of aquaculture and the challenge of bacterial diseases in freshwater farmed fish in Tanzania; A call for sustainable strategies. Aquaculture and Fisheries 6(3): 247-253.

Niyibizi, L., A. Vidakovic, A. Norman Haldén, S. Rukera Tabaro and T. Lundh (2022). Aquaculture and aquafeed in Rwanda: current status and perspectives. Journal of Applied Aquaculture 35(3): 743-764.

Nobile, A. B., A. M. Cunico, J. R. S. Vitule, J. Queiroz, A. P. Vidotto□Magnoni, D. A. Z. Garcia, M. L. Orsi, F. P. Lima, A. A. Acosta, R. J. da Silva, F. D. do Prado, F. Porto□Foresti, H. Brandão, F. Foresti, C. Oliveira and I. P. Ramos (2019). Status and recommendations for sustainable freshwater aquaculture in Brazil. Reviews in Aquaculture 12(3): 1495-1517.

Nowland, S. J., W. A. O'Connor, M. W. J. Osborne and P. C. Southgate (2019). Current Status and Potential of Tropical Rock Oyster Aquaculture. Reviews in Fisheries Science Aquaculture 28(1): 57-70.

O'Donncha, F. and J. Grant (2019). Precision Aquaculture. IEEE Internet of Things Magazine 2(4): 26-30.

Ochieng Ogello, E. and J. M. Munguti (2016). Aquaculture: a promising solution for food insecurity, poverty and malnutrition in Kenya. African Journal of Food, Agriculture, Nutrition and Development 16(4): 11331-11350.

Pramulati, I., Hartaty, H., Sukmaningsih, A. A. S. A., and Sudaryanto, F. X. (2023). Length at first maturity and length at first capture for bullet tuna (Auxis rochei Risso, 1810) in the southern waters of Bali. In BIO Web of Conferences (Vol. 74, p. 03008). EDP Sciences.

Rana KJ, Hasan MR (2013). On-farm feeding and feed management practices for sustainable aquaculture production: an analysis of case studies from selected Asian and African countries. In: Hasan MR, New MB (eds) On-farm feeding and feed management in aquaculture. Rome: FAO Fisheries and Aquaculture Technical Paper 583:21-67.

Satia, B. P., Anozie, O., Seisay, M., and Nouala, S. (2016). Implementation of international fisheries instruments in Africa: A case study for Central Africa.

Satia P. B. (2017). Regional review on status and trends in aquaculture development in sub-Saharan Africa – 2015, FAO Fisheries and

Schmidt, C. C. (2003). Globalisation, industry structure, market power and impact of fish trade. Opportunities and challenges for developed (OECD) countries. Report of the Expert Consultation on International Fish Trade..

Shen, Y., K. Ma and G. H. Yue (2021). Status, challenges and trends of aquaculture in Singapore. Aquaculture 533.

Valenti, W. C., H. P. Barros, P. Moraes-Valenti, G. W. Bueno and R. O. Cavalli (2021). Aquaculture in Brazil: past, present and future. Aquaculture Reports 19.

Vince, J., B. D. Hardesty and C. Wilcox (2020). Progress and challenges in eliminating illegal fishing. Fish and Fisheries 22(3): 518-531.

3

The Future of Aquaculture Prospects and Challenges

***Akhilesh Kushwaha*[1], *Monika Maurya*[2] and *Shailja Sen*[3*]**

[1]*Department of Horticulture, Sam Higginbottom University of Agriculture Technology and Sciences, Prayagraj-211007, Uttar Pradesh, India*

[2]*Department of Biological Sciences, Sam Higginbottom University of Agriculture Technology and Sciences, Prayagraj-211007, Uttar Pradesh, India*

[3]*Sam Higginbottom University of Agriculture Technology and Sciences Prayagraj-211007, Uttar Pradesh, India*

Abstract

Aquaculture, the farming of aquatic organisms, holds immense promise for meeting the growing global demand for seafood while alleviating pressure on wild fish stocks. This abstract explores the future trajectory of aquaculture, highlighting both its prospects and the challenges it faces. The future of aquaculture appears bright, with projections indicating continued growth in production to meet rising demand for seafood. Advancements in technology, such as recirculating aquaculture systems (RAS) and genetic improvements in fish breeds, offer opportunities for increased efficiency, productivity, and sustainability. Furthermore, the diversification of species farmed, including high-value marine species and alternative protein sources such as algae and insect-based feeds, opens new avenues for innovation and market expansion. Despite its potential, aquaculture confronts numerous challenges that must be addressed to realize its full benefits sustainably. Environmental concerns, such as habitat degradation, water pollution, and disease transmission, remain significant challenges requiring robust management strategies. Additionally, social and economic considerations, including equitable access to resources, labor rights, and market dynamics, present complex obstacles to the industry's growth and inclusivity. Furthermore, regulatory frameworks must adapt to evolving technologies and practices to ensure responsible and ethical aquaculture development.

1. Introduction

What is aquaculture?

The term aquaculture broadly refers to the cultivation of aquatic organisms in controlled aquatic environments for any commercial, recreational or public purpose. The breeding, rearing and harvesting of plants and animals takes place in all types of water environments including ponds, rivers, lakes, the ocean and man-made "closed" systems on land.

OR

Aquaculture is the breeding, rearing, and harvesting of fish, shellfish, algae, and other organisms in all types of water environments. As the demand for seafood has increased, technology has made it possible to grow food in coastal marine waters and the open ocean. Aquaculture is a method used to produce food and other commercial products, restore habitat and replenish wild stocks, and rebuild populations of threatened and endangered species.

There are two main types of aquacultures—marine and freshwater. NOAA efforts primarily focus on marine aquaculture, which refers to farming species that live in the ocean and estuaries.

In the United States, marine aquaculture produces numerous species including oysters, clams, mussels, shrimp, seaweeds, and fish such as salmon, black sea bass, sablefish, yellowtail, and pompano. There are many ways to farm marine shellfish, including "seeding" small shellfish on the seafloor or by growing them in bottom or floating cages. Marine fish farming is typically done in net pens in the water or in tanks on land. U.S. freshwater aquaculture produces species such as catfish and trout. Freshwater aquaculture primarily takes place in ponds or other manmade systems. NOAA is committed to supporting an aquaculture industry that is economically, environmentally and socially sustainable. NOAA experts and partners work to understand the environmental effects of aquaculture in different settings and provide best management practices to help reduce the risk of negative impacts.

Fig. 1. Farm showing Aquaculture Setup.

1.1 What is Aquaculture and Why is it Important?

Because 70% of the planet's surface is covered by water, humans have understood its importance as a resource. This is why aquaculture is the most exploited in terms of water use as a resource. This is especially true in food production as opposed to land use.

It fulfills various purposes, including:

- food production
- restoring populations that are threatened and endangered
- population enhancement of wild stocks
- aquarium construction
- fish farming
- Home rehabilitation

1.1.1 Examples of aquaculture

Mariculture is the branch of aquaculture that cultivates marine organisms. It is practiced either:

- at sea
- in a closed part of the ocean
- in basins or ponds filled with sea water.

Organisms grown in salt water are:

- fish like flounder and whiting
- Seafood (such as shrimp and oysters)
- marine plants (such as kelp and seaweed)

Cosmetics and jewelry such as cultured pearls also use the products of mariculture. Algaculture is the type of aquaculture that cultivates seaweed. Most of the algae collected are microalgae (phytoplankton, microphytes or planktonic algae) or macroalgae, commonly known as seaweed. Although macro-algae are used for various commercial purposes, its size and cultivation requirements make it difficult to grow. Microalgae are easier to grow on a large scale.

2. Plants in aquaculture

Aquaculture refers to the cultivation of aquatic organisms in a safe and controlled way. It uses several types of waste water as a source of nutrients and/or warm temperatures for the growth of plants and fish. In the production of aquaculture plants, there are two main methods of cultivation:

2.1 Hydroponics: the place where plants are cultivated with roots directly exposed to water

2.2 Floating Plant Ponds: A floating plant pond is a modified maturation pond with floating plants (macrophytes). Plants such as hyacinths or duckweed float in them while the roots hang in the water to absorb nutrients and filter the flowing water.

3. Aquaculture History

The history of aquaculture started in China around 3000 BC. Artificial lakes contained specific fish such as carp when the water level was lowered after river floods. Their eggs were then fed with nymphs and excrement of silkworms used for silk production.

The romans were quite skilled at breeding fish in ponds. In Europe, fish was rare and therefore expensive and became common in monasteries in the Middle Ages. 19th century improvements in transportation made fish readily available and inexpensive, even far from the seas, leading to a decline in aquaculture.

The current boom began in the 1960s after overfishing caused prices to rise again. Today, commercial aquaculture exists on a considerable scale that was previously unheard of, generating controversy because of its effects on public waters beyond enclosure boundaries.

3.1 Aquaculture production numbers

Global fish production reached 171 million tons in 2016, with aquaculture accounting for 47% of the total and 53% if uses are not excluded. The total value of the first sale of fisheries and aquaculture production in 2016 was estimated at 362 billion USD. And including 232 billion USD from aquaculture production. With capture fishery production relatively stable since the late

1980s, aquaculture has been responsible for the impressive and continued growth in the supply of fish for human consumption.

Global aquaculture production in 2016 included:

- 80.0 million tons of food fish
- 30.1 million tons of aquatic plants
- As well as 37,900 tons of non-food products.

Farmed food fish production includes:

- 54.1 million tons of fish,
- 17.1 million tons of shellfish
- 7.9 million tons of shellfish
- 938,500 tons of other aquatic animals

China, because of its history of aquaculture, has produced more than the rest of the world every year since 1991 in 2016. Other top producers in 2016 were India, Indonesia, Viet Nam, Bangladesh, Egypt and Norway.

4. Types of Aquacultures

Several aquaculture practices are applied worldwide. There are indeed several types of aquacultures:

- extensive or intensive
- in a natural environment or in a pond
- in fresh or salt water
- in circulation or recirculation systems

If these practices allow a great diversity of breeding organisms, they require specific techniques and equipment.

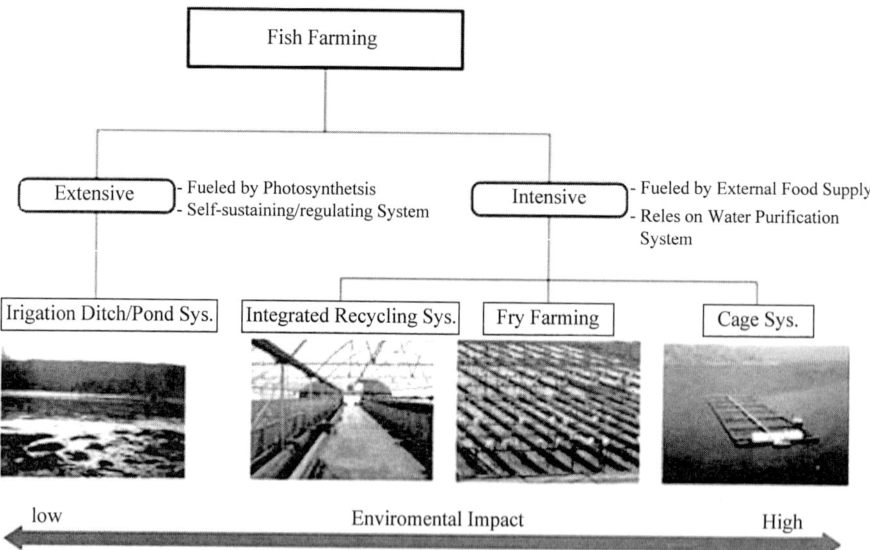

Fig. 2. Picture showing different types of aquaculture.

4.1 Aquaculture Engineering

At the outlet of the fish tank, the water first flows to a mechanical filter allowing the elimination of the contained organic waste. The finest particles and dissolved compounds such as phosphate then pass through the mechanical filter.

This is why the water goes to a biological filter, to decompose organic matter and ammonia. All this, before being aerated and cleaned of carbon dioxide and returned to the aquariums.

This is the biological process carried out by bacteria in the bio filter.

4.1.1 Fish Farming Ponds: In the water, fish feed and produce excrement, rich in ammonia.

4.1.2 Mechanical Filtration: The particulate and dissolved waste contained in the water of the rearing tanks is eliminated through a drum filter.

4.1.3 Water Flow Pond: The water flows to a first intermediate storage pond, the water flow pond.

4.1.4 Uv Disinfection: An ultraviolet treatment destroys the bacteria and viruses contained in the water coming from the breeding ponds.

4.1.5 Biological Filtration: A biological filter supports specific bacteria that convert ammonia into non-toxic nitrate.

5. Recirculating aquaculture (RAS)

In an aquaculture system recirculation, on purified reutilize l'eau de culture en permanence. Indeed, a recirculating aquaculture system is practically a closed circuit.

- The waste products produced; solid waste, ammonium and CO2, are either removed or converted to non-toxic products by the system components.
- The purified water is then saturated with oxygen and returned to the aquariums.
- By recirculating crop water, water and energy requirements are kept to a minimum. However, it is not possible to design a completely closed recirculation system.
- Non-degradable waste should be removed and evaporated water replaced.

Nevertheless, our recirculating systems are capable of reusing 90% or more of the crop water. To ensure proper water purification, we provide recirculating systems with a number of components with specific functions. The benefits of farming in RAS are:

- Fully controlled environment for fish
- Low water consumption
- Efficient energy use
- Efficient land use
- Optimum feeding strategy
- Easy grading and harvesting of fish
- Total control of disease
- Recirculating aquaculture systems constraints

The growing aquaculture recirculation systems are the best option for sites close to or in cities, with good availability of electricity. Furthermore, the use of RAS technology is the only possibility for rearing tropical fish species in moderate to cold indoor climates.

6. What is fish farming

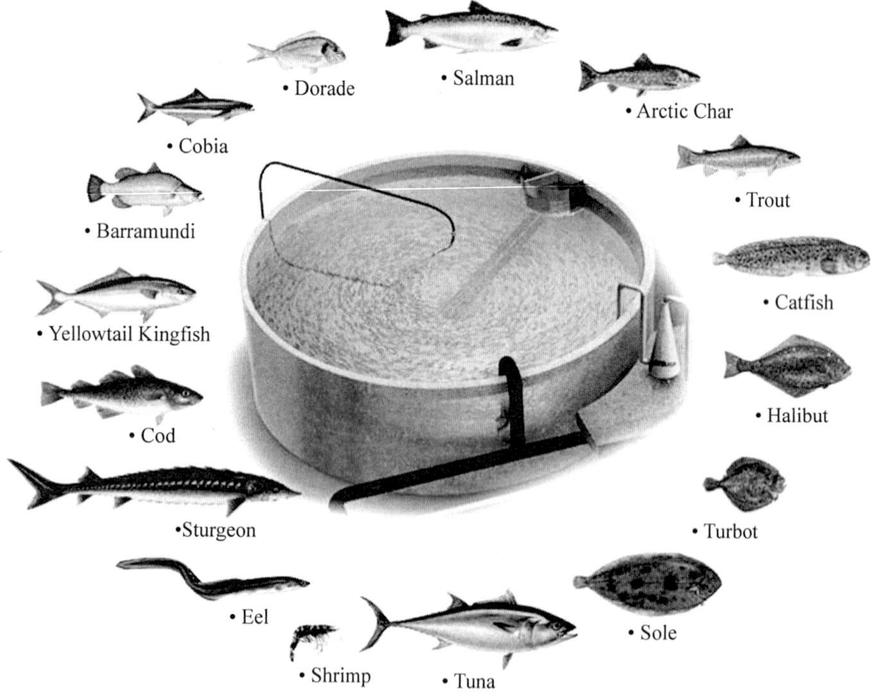

Fig. 3. Different types of fish used in Aquaculture.

Fish farming is one of the branches of aquaculture. It involves the commercial rearing of fish in tanks or pens such as fish ponds, usually for food. This is the principal method of aquaculture, while other methods can be assimilated to mariculture. A structure that releases wild juvenile fish for sport fishing or to supplement the natural numbers of a species is generally referred to as a hatchery. Fish farming professionals are grouped into unions or national representative bodies. Indeed, they carry out their activity within the framework of a respectful approach of the environment while focusing on a high-quality production. There are 33.8 million tons worth about US$60 billion according to the 2008 global fish farming revenues recorded by the FAO. This makes fish farming a sector with great economic potential.

6.1 Fish farming process

6.1.1 Method 1

The first method is through a system of cages placed in lakes, ponds and oceans containing the fish. This method is also commonly referred to as sea farming. Fish are kept in cages as part of the structure and are "artificially fed" and harvested. This cage method has made many technological advances over the years, especially with the reduction of disease and environmental concerns. However, the main concern with this method is that the fish escape and become detached among the wild fish population.

6.1.2 Method 2

The second method is to use ditch or irrigation pond systems to raise fish. The basic requirement for this method is to have a ditch or pond that holds water.

This system is unique because:

On a small basis, the fish are fed artificially and the waste produced by the fish is then used to fertilize the farmers' lands.

On a larger scale, especially in ponds, the pond is self-sufficient because it produces plants and algae to feed the fish.

6.1.3 Method 3

The third method of fish farming is composite fish farming. This is a type that allows local and imported fish species to coexist in the same pond. The number of species may depend, but it is possible to have more than six species of fish in a single pond. Fish species are always carefully selected to ensure coexistence and reduce competition for food.

6.1.4 Method 4

This method is considered to be the most common "pure" method of fish farming.

- This approach uses large plastic tanks placed in a greenhouse.
- There are hydroponic beds placed near the plastic tanks.
- The water in the clay tanks is pumped to the hydroponic beds
- This is where fish waste feed is used to provide nutrients to plants grown in hydroponic beds. The majority of the types of plants grown in hydroponic beds are herbs such as parsley and basil.

The last type of fish farming method is the "continuous flow system". This is when farmed fish species are raised from eggs and then dumped into streams.

7. Fish Farming Project

Pond fish farming, with an earthen basin, in which the fish feed themselves completely or partially from the biological production of the environment.

Intensive fish farming in artificial ponds or cages, in which the fish are fed exclusively with food provided by the fish farmer.

7.1 Fish used for fish farming

For the fish farming technique, different types of freshwater and saltwater fish can be produced. Freshwater fish species are:

- The Rainbow Trout is undoubtedly the most widespread. Rainbow trout is the most produced species with over 32,000 tons.
- The Fario trout. Little tamed, it is essentially intended for the repopulation of rivers.
- Brook Trout or Brook Salmon
- Arctic Char
- Sturgeon
- Carp

7.2 Benefits of Aquaculture

Aquaculture offers alternatives to marine fisheries. Increased demand for food products and globalization has led to an increase in fishing. However, this has led to fishermen becoming selfish and overfishing on desirable or high demand species. With aquaculture, they offer both an alternative and an opportunity for wild stocks to replenish their overtime.

7.3 Aquaculture Efficiency

Fish convert nutrients into body protein more efficiently than cattle or chickens. This means that fish companies are producing more feed for less feed. Such efficiency means that less food and energy are used to produce food. Moreover, the manufacturing process is also less expensive. It saves resources and even produces more food. This allows to secure the reserves and minimize the environmental constraints.

8. Economic Impact

8.1 Dependence on aquaculture

Fish and other seafood are good sources of protein. They also have a higher nutritional value, such as the addition of natural oils in the diet, such as omega-3 fatty acids. Also, since it provides white meat, it is better for cholesterol reduction in the blood than red beef. Fish are also easier to keep than other meat-producing animals because they are able to convert more nutrients into protein. Therefore, his overall conversion of pounds of feed to pounds of protein makes it less expensive to raise fish because they use it more efficiently. Sea weeds are progressively transformed into alternative fuel sources by making them produce fuels that can replace contemporary fossil fuels. Algae produce lipids that, if harvested, can be burned as an alternative fuel source, the only byproduct of which would be water when burned.

Such a progress could mitigate the world's dependence on drilled fossil fuels as well as reduce the price of energy by growing it instead of drilling for oil. In addition, seaweed-based fuel is a cleaner, workable energy source, which means it can help revolutionize the energy sector and create a more stable economy that avoids the explosive nature of oil and replaces it with a more abundant fuel source.

9. Purpose of Aquaculture

Aquaculture serves many purposes, including:

- Food production for human consumption
- Rebuilding of populations of threatened and endangered species
- Habitat restoration
- Wild stock enhancement
- Production of baitfish
- Fish culture for zoos and aquariums

It is one of the fastest growing forms of food production in the world. Because harvest from many wild fisheries has peaked globally, aquaculture is widely recognized as an effective way to meet the seafood demands of a growing population. Using aquaculture techniques and technologies, researchers and the aquaculture industry are "farming" all types of freshwater and marine species of fish and shellfish:

9.1 Marine aquaculture refers specifically to the culturing of oceanic species (as opposed to freshwater). Examples of marine aquaculture production include oysters, clams, mussels, shrimp, salmon and algae. Marine aquaculture is just 20 percent of U.S. production, consisting mostly of shellfish (e.g., oysters, clams and mussels).

9.2 Freshwater aquaculture includes trout, catfish and tilapia. About 70 percent of aquaculture in the United States is freshwater farming of catfish and trout. Only a handful of U.S. farms grow marine finfish such as salmon in Maine and Washington State and yellowtail and Pacific threadfin (Moi) in Hawaii. Aquaculture produces almost half of the seafood consumed by humans globally, a trend that continues to increase. The United States is a major consumer of aquaculture products — we import 84 percent of our seafood, and half of that is from aquaculture — yet we are a minor producer. In fact, U.S. aquaculture (both freshwater and marine) supplies about 5 percent of the U.S. seafood supply, and U.S. marine aquaculture supplies less than 1.5 percent.

It is vital that the United States further develop its own sustainable aquaculture industry, both to reduce its annual $9 billion annual seafood import deficit and to keep pace with the growing demand for seafood.

10. The Future of Aquaculture In India

Aquaculture, the farming of aquatic organisms, is emerging as a significant contributor to global seafood production, addressing the increasing demand for fish and other aquatic products. In India, with its vast coastline and abundant water resources, aquaculture presents immense opportunities for sustainable development, food security, employment generation, and economic growth. This article examines the future of aquaculture in India, highlighting the government's role, current trends, and potential challenges.

11. Current State of Aquaculture in India

Aquaculture in India has witnessed remarkable growth over the past few decades. According to the Central Marine Fisheries Research Institute (CMFRI), the total fish production in India was approximately 13.78 million metric tons in 2019-2020, out of which aquaculture contributed around 6.46 million metric tons (47%). It is estimated that by 2030, aquaculture will account for over 70% of the total fish production in India.

12. Government Initiatives and Policies

The Government of India recognizes the importance of aquaculture and has implemented various initiatives and policies to promote its sustainable growth. The key government initiatives include:

12.1 National Fisheries Development Board (NFDB): NFDB is responsible for providing technical assistance, financial support, and capacity building to the aquaculture sector. It focuses on enhancing productivity, promoting research and development, and ensuring sustainable aquaculture practices.

12.2 Pradhan Mantri Matsya Sampada Yojana (PMMSY): Launched in 2020, PMMSY is a flagship scheme aimed at boosting fisheries and aquaculture infrastructure, improving post-harvest management, and enhancing value addition. It targets doubling fishers' incomes and attracting private investment in the sector.

12.3 Coastal Aquaculture Authority (CAA): CAA regulates and monitors coastal aquaculture activities, ensuring environmentally sustainable practices and preventing unauthorized activities. It plays a crucial role in maintaining the ecological balance of coastal regions.

12.4 Marine Products Export Development Authority (MPEDA): MPEDA focuses on promoting aquaculture exports, providing market access, and supporting aquaculture farmers in adopting quality control measures. It facilitates the development of infrastructure and the implementation of international quality standards.

13. Potential for Growth

India's diverse climatic conditions, vast coastline, and inland water resources provide immense potential for the growth of aquaculture. The sector offers significant opportunities for small-scale farmers, entrepreneurs, and coastal communities. Key factors contributing to the growth potential include:

13.1 Rising Domestic and Global Demand: The increasing population, changing dietary preferences, and growing health consciousness are driving the demand for fish and seafood products. Additionally, India's growing export market presents significant opportunities for aquaculture farmers.

13.2 Technological Advancements: Adoption of advanced technologies, such as recirculating aquaculture systems (RAS), biofloctechnology, and genetic improvement programs, can enhance productivity, improve resource utilization, and mitigate environmental impacts.

13.3 Diversification of Species: While carps and shrimps dominate Indian aquaculture, diversifying into high-value species like seabass, grouper, and tilapia presents opportunities for premium markets and higher profits.

13.4 Inland Aquaculture Potential: India's large reservoirs, ponds, and wetlands offer immense potential for inland aquaculture, particularly in states like Andhra Pradesh, West Bengal, and Odisha.

14. Challenges And Sustanibility Considration

Despite the promising future, aquaculture in India faces certain challenges that need to be addressed for sustainable growth:

14.1 Environmental Concerns: Intensive aquaculture practices can lead to water pollution, habitat degradation, and disease outbreaks. Implementing responsible aquaculture practices, minimizing chemical usage, and adopting waste management strategies are crucial for mitigating these impacts.

14.2 Disease Management: The aquaculture sector is vulnerable to diseases, which can lead to substantial economic losses. Investing in disease surveillance, developing disease-resistant strains, and promoting biosecurity measures are essential for minimizing disease outbreaks.

14.3 Access to Finance and Technology: Small-scale farmers often face challenges in accessing finance and adopting advanced technologies.

Encouraging financial institutions to provide tailored financial products and promoting technology transfer through public-private partnerships can address these barriers.

15. Advantages and Disavantages of Aquaculture

The following are the advantages and disadvantages of Aquaculture:

15.1 Advantages of Aquaculture

15.1.1 Boosts food production – Aquaculture helps to increase the production of seafood, meeting the growing demand for food worldwide. It's a great way to ensure a steady supply of fish and other seafood.

15.1.2 Creates job opportunities – It also opens up job opportunities. Many people can find work in fish farming, from the actual farming to the selling of the produce.

15.1.3 Conserves wild fish populations – Wild fish populations are under threat from overfishing. Aquaculture can help conserve these populations by reducing the need to catch wild fish.

15.1.4 Supports economic growth – Aquaculture can contribute to a country's economic growth. It can bring in money from the sale of fish and other seafood, both locally and internationally.

15.1.5 Enhances nutrient recycling – Aquaculture also plays a role in nutrient recycling. The waste produced by fish can be used as a nutrient source for plants, helping to create a sustainable ecosystem.

15.2 Disadvantages of Aquaculture

15.2.1 Can harm natural ecosystems – Aquaculture can damage natural ecosystems by introducing non-native species that alter the balance of life.

15.2.2 Overuse of antibiotics – It often involves the overuse of antibiotics, which can lead to resistant bacteria and harm human health.

15.2.3 Risk of disease spread – There's a risk of disease spread among farmed fish, which can then infect wild populations.

15.2.4 Depletes wild fish populations – The practice can deplete wild fish populations, as they are often used to feed farmed fish, disrupting the food chain.

16. Challenges and Vision

In June 2018, GSI and the World Wildlife Fund (WWF) co-hosted a thought-leaders' discussion at the World Bank to examine the Future of Aquaculture. Focusing on questions such as: what barriers and opportunities might the industry face? And what can we do now to help ensure the sector's future is sustainable? The group set out its vision for the future and identified a number of change levers which it sees as vital in helping the industry achieve this vision.

16.1 The Challenges

Population growth, aging populations, increases in income and wealth, as well as healthier diets are all leading to increased seafood consumption. However, we cannot extract more seafood from our oceans, which is why sustainable growth in aquaculture is needed.

16.2 The Vision

Aquaculture has significant potential in helping provide a healthy and sustainable protein source for future populations. However, to reach this potential, an increase in production is needed. This increased production must be responsibly managed and matched with significant reductions in environmental impact and improvements in resource efficiency.

16.2.1 4 ways to ensure the future of aquaculture is sustainable

- Demand for blue foods (food from aquatic sources) is expected to double by 2050.
- Aquaculture will play a key role in boosting supplies of nutritious and healthy food for billions of consumers around the world.
- Stakeholder partnerships need to be developed to ensure aquaculture becomes a more sustainable and climate-friendly industry.
- As our global population expands, so too does demand for nutritious and climate-friendly food. Meeting this increased appetite in more sustainable ways represents a monumental challenge, but potential solutions lie in the planet's waters.

Blue foods, which are sourced from oceans, seas, rivers and lakes, are the most highly traded food products in the world. Not only do these foods provide livelihoods for millions of people, they also feed billions every day. Demand for blue foods is expected to double by 2050, with aquaculture production playing a vital role in augmenting supplies. In fact, the UN's Food and Agriculture Organization anticipates that aquaculture will continue to drive growth in global fish production by accounting for 106 million metric tons in 2030 - a rise of 32% from 2020 levels.

Conclusion

The future of aquaculture in India holds great promise for sustainable development, food security, and economic growth. With robust government initiatives, technological advancements, and a vast resource base, India can position itself as a global aquaculture powerhouse. However, addressing environmental concerns, ensuring disease management, and promoting inclusive growth will be crucial for realizing the full potential of the sector.

References

Arthur, R.I., Lorenzen, K., Homekingkeo, P., et al., 2010. Assessing impacts of introduced aquaculture species on native fish communities: Nile tilapia and major carps in SE Asian freshwaters. Aquaculture 299, 81–88.

Bagarinao, T.U., Primavera, J.H., 2005. Code of Practice for Sustainable Use of Mangrove Ecosystems for Aquaculture in Southeast Asia. Iloilo, Philippines: SEAFDEC Aquaculture Department.

Belak, J., Dhar, A.K., Primavera, J.H., et al., 1999. Prevalence of viral diseases (IHHNV and WSSV) in Penaeus monodon from the Philippines andits association with mangrove status and shrimp culture systems. In: Alcivar-Warren, A. (Ed.), Proceedings of the Aquaculture and Conservation of Marine Shrimp Biodiversity Symposium. North Grafton, MA: Tufts University School of Veterinary Medicine.

Beveridge, M.C.M., Phillips, M.J., Dugan, P., Brummett, R., 2010. Barriers to aquaculture development as a pathway to poverty alleviation and food security. In: OECD (Ed.), Advancing the Aquaculture Agenda: Workshop Proceedings. Paris: OECD Publishing, pp. 345–359.

Chopin, T., Robinson, S.M.C., Troell, M., et al., 2008. Ecological engineering: Multi-trophic integration for sustainable marine aquaculture. In: Jorgensen, S.E., Fath, B. (Eds.), Encyclopedia of Ecology. Amsterdam: Elsevier, pp. 2463–2475.

Henriksson, P., Pelletier, N., Troell, M., Tyedmers, P., 2012. Life cycle assessment and its application to aquaculture production systems. In: Meyers, R.A. (Ed.), Encyclopedia of Sustainability Science and Technology, first ed. New York: Springer-Verlag.

Huntington, T.C., Hasan, M.R., 2009. Fish as feed inputs for aquaculture – Practices, sustainability and implications: A global synthesis. In: Hasan, M.R., Halwart, M. (Eds.), Fish as Feed Inputs for Aquaculture: Practices, Sustainability and Implications. Rome: FAO, pp. 1–61.

Verdegem, M.C.J., Bosma, R.H., Verreth, J.A.V., 2006. Reducing water use for animal production through aquaculture. Water Resources Development 22, 101–113.

4

Aquaculture and Climate Change

Khusbu Samal, Narendra Kumar Maurya* and Bhooleshwari

College of Fisheries, Mangalore, Karnataka Veterinary, Animal and Fisheries Sciences University, Bidar-575002, Karnataka, India

Abstract

Aquaculture is now the food production industry with the greatest rate of growth in the world due to its continued enormous output expansion. However, the sector's sustainability is in jeopardy because of the anticipated impacts of climate change, which are both a present and a future reality. In addition to the numerous stresses that fish stocks currently face-such as overfishing, habitat loss, pollution, disturbances, and invasive species-climate change is an extra one. It is necessary to include other anthropogenic forces when assessing the effects of climate change. The increasing frequency and intensity of extreme weather events can have significant consequences, such as storms that destroy property or flood freshwater fields. Different physiological impacts and pressures will alter the growth and development of fish and shellfish, thereby making them more vulnerable to infections and illnesses. In this research, we examine how climate change may affect aquaculture productivity and how that may affect the sustainability of the industry. There has been discussion on several aspects of a changing climate, including increasing temperatures, sea level rise, illnesses and toxic algal blooms, altered rainfall patterns, the unpredictability of outside input sources, altered sea surface salinity, and extreme weather events.

Keywords: *Aquaculture, Climate Change, Ocean Acidification and Rising Temperature*

1. Introduction

Aquaculture production is impacted by climate change in both direct and indirect ways. Indirect effects can arise through changing the primary and secondary productivity, ecosystem structure, input supplies, or the prices of products, fishmeal and fish oil, and other goods and services required by fishers and aquaculture producers. Direct effects can also include influencing the physical state and physiology of finfish and shellfish stocks in production systems.

The original book goes into great length about the ways in which aquaculture production will be impacted by climate change and the implications this will have for the sector's sustainability.

Aquaculture production is interconnected with other food production systems and does not happen in a vacuum. Scholars have noted that in order to meet the growing demand for aquatic goods in a sustainable manner, it is important to acknowledge the close relationship that exists between the objectives of aquaculture, agricultural, and fisheries systems.

Global food production is currently seen to be at risk from climate change, which also poses a serious threat to both the quantity and quality of production. The anticipated effects of climate change are posing a growing danger to food security, namely to access to dietary protein. A lot of research has been done on the implications of climate change on aquaculture, both locally and globally, due to the industry's major contribution to livelihoods, nutrition, and food security worldwide.

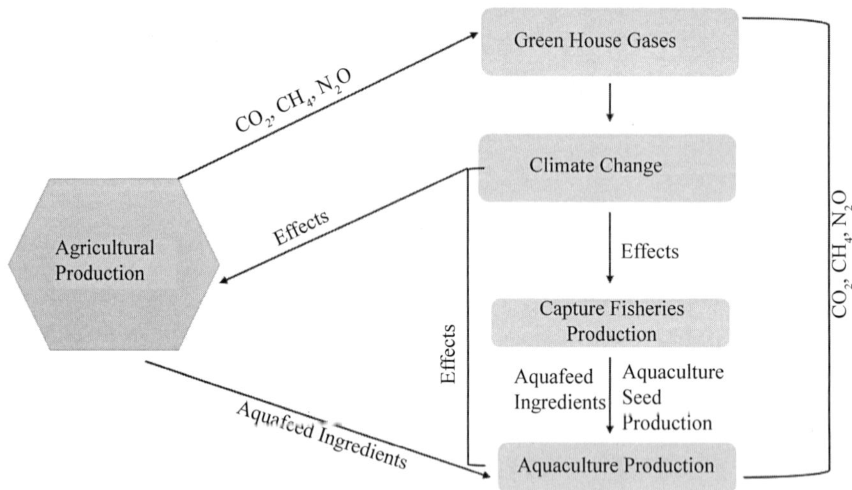

A simple illustration of the direct and indirect pathways through which climate change will affect aquaculture production. An illustration of how GHG emissions will affect aquaculture production as well as the contribution of capture fisheries, aquaculture, and agriculture activities to climate change.

2. Primary issues brought on by climate change making Aquaculture Vulnerable

2.1 Rising temp

For aquatic species to grow and thrive, temperature is essential (Ngoan, 2018). Because they are poikilothermic, fish may be especially vulnerable to changes in temperature brought on by climate change (Sae-Lim *et al.*, 2017; Adhikari *et al.*, 2018). The majority of fish, particularly cold-water species like Atlantic halibut, salmon, and cod, as well as intertidal shellfish, are expected to experience higher mortality rates due to thermal stress as a result of the anticipated 1.5°C increase in the average world temperature this century. As a result, extended heat stress may have a variety of effects on aquaculture productivity, the main one being decreased output. Chronic stress, for instance, may impact the neuroendocrine and osmoregulatory systems, changing the capacity for aerobic respiration and performance, as well as the immunological responses of a number of commercially significant species. Additionally, most finfish and shellfish species' metabolism, physiology, eating habits, and growth performance are probably going to be impacted. Furthermore, the ocean's ability to absorb carbon dioxide is gradually diminished by rising ocean temperatures and the ensuing ocean acidification, which results in changes to water systems' hydrology and hydrography as well as the occurrence of red tides (Cochrane *et al.*, 2009). The economic and social viability of aquaculture production may be threatened by these impacts, which could also result in low productivity and higher management expenses. Thermal stratification in deep water bodies brought on by temperature fluctuations may also have an impact on environmental sustainability. This is because it can alter the amount and distribution of nutrients in the water as well as aquaculture in the event of an upwelling. The production of warmer water species, such as the giant tiger prawn, tilapia, oysters, and mussels, may be favoured by warmer periods (within the tolerance conditions of the species), particularly in temperate regions. The decline of desired specimens in the wild due to eroding coral reefs may present chances for larger-scale investors who operate hatcheries in protected places. Warmer stretches are anticipated to encourage the growth of aquaculture output in colder places, such the Arctic. Additionally, warmer seasons can present chances to cultivate novel species and support ongoing advancements in aquatic organisms' genetic modifications. Through higher production outputs and job possibilities, these prospects will support social sustainability; similarly, they will support economic sustainability through higher profits and lower management costs in these domains. To do this, though, will need the development of molecular biology and the application of useful genetic improvement techniques in aquaculture, which could jeopardize environmental sustainability in the event that native species hybridize.

2.2 Ocean acidification

Ocean acidification is the outcome of atmospheric CO2 uptake causing a prolonged (often decades) drop in the pH of ocean water. According to scientists, the oceans are thought to hold 50 times more CO2 than the atmosphere. Many aquatic species' growth, development, calcification, survival, and abundance will be negatively impacted by the anticipated increase in CO2 uptake by seas at 1.5°C or higher global warming (IPCC, 2018). An increase in CO2 buildup in water may cause a decrease in pH and an increase in acidity, which could endanger the environmental sustainability of aquaculture production systems by degrading water quality and resulting in low productivity. Furthermore, increasing ocean acidity decreases the availability of carbonate needed by shell-forming organisms like shrimp, mussels, oysters, or corals to form coral skeletons (Calcification). This could pose a threat to the production of marine aquaculture. For instance, poor coral skeleton formation may lead to increased juvenile predation rates, which in turn reduces collection rates and reduces wild spat oyster production (Blanchard *et al.*, 2017). Large-scale producers will therefore probably be forced to raise their production costs by depending on hatcheries to generate spat oysters. The social and economic sustainability of aquaculture production in these areas may be negatively impacted, as well as production outputs and revenues. By upsetting the intercellular transport processes in saltwater, increasing acidity levels could have a substantial impact on aquatic organisms' physiology and metabolism. The ability of species to adapt to changes in ocean acidity will depend on a variety of factors, including complex biophysical feedbacks, species adaptive capacity, and the rate of change. This ability to adapt is difficult to capture at a global scale. Macroalgal (seaweed) production may also be affected by ocean acidification, but such effects would depend on the acquisition kinetics of inorganic carbon by different species. The fact that large-scale aquaculture producers may become entirely or partially dependent on hatcheries for the generation of spats could have a positive effect on the hatchery owners' bottom line. Furthermore, as these hatcheries will need a lot of personnel to meet the increasing demand, this might lead to additional job possibilities for the local community and support the social and economic sustainability of aquaculture production. Finfish, especially marine species, are susceptible to ocean acidification due to the presence of calcified otolithsdsr. This may have an impact on their growth and development (Frommel *et al.*, 2014), RNA viability, damage to tissues, and respiration. Despite this, the effects of ocean acidification on finfish are not well understood.

2.3 Diseases and harmful algal blooms

Finfish, especially marine species, are susceptible to ocean acidification due to the presence of calcified otoliths. This may have an impact on their growth and development (Frommel *et al.*, 2014), RNA viability (Franke and Clemmesen, 2011), damage to tissues, and respiration (Munday *et al.*, 2009; Frommel *et al.*, 2012). Despite this, the effects of ocean acidification on finfish are not well understood (Wittmann and Portner, 2013; Clements and Chopin, 2016). Therefore, it is anticipated that as a result of climate change, outbreaks of warm water diseases will happen more frequently in addition to the potential for the discovery of new ones (Sae-Lim *et al.*, 2017). A number of finfish and shellfish species are expected to have faster pathogen replication, pathogenicity, longer life cycles, and transmission as a result of rising temperatures. Furthermore, the strain of rising temperatures may encourage the emergence of epizootic diseases in aquaculture and result in significant economic difficulties. According to Maulu *et al.*, (2019), the breakout of epizootic illnesses continues to be a significant factor limiting the viability of aquaculture production systems in many countries across the world. Additionally, it is anticipated that warm water diseases, such sea lice, would continue to be a problem for salmon aquaculture, and that additional warming will probably make infections in colder climates worse, necessitating more frequent treatments and higher costs. Overall, a higher prevalence of diseases in aquaculture production systems will result in lower revenues, which will have an impact on the aquaculture production's social and economic sustainability. On the other hand, the emergence of unfavourable conditions owing to vibriosis and winter ulcer, two illnesses that harm Atlantic salmon, may cause them to progressively go extinct. This could be advantageous for the growth of this particular fish species.

2.4 Changes in rainfall pattern

Rainfall patterns that shift will have two directly opposing effects on aquaculture productivity and sustainability: periods of high or no rainfall (called flooding) and periods of low or no rainfall (called drought). The IPCC (2018) states that while patterns of flooding occurrences are hard to anticipate with precision, hazards associated with drought events are projected to be increased at 2°C compared with 1.5°C of global warming in a given region. Raising rainfall amounts will raise the hazards to output in lowland areas, especially if they coincide with larger occurrences. These hazards include fish loss from ponds during floods, undesired species invading the pond, and wall erosion and infilling causing damage to the pond. The introduction of invasive fish species and degradation of water quality are the main ways in which the mixing of pond water and fish with wild fish could have a detrimental impact on the environmental sustainability of aquaculture output. Additionally, because fish losses from ponds reduce producer profits and spread poverty throughout communities, they pose a danger to the social and economic aspects of aquaculture sustainability. However, Rutkayova *et al.*, (2017) state that the age and species of each particular species will determine the percentage of fish lost during periods of significant floods. The scientists also noted that as fish get older, the percentage of losses should go down. It should be highlighted, nonetheless, that more rainfall may expand the areas in low-lying tropical regions that are appropriate for rainwater-dependent aquaculture ponds, supporting the social and economic viability of those places. Additionally, it has been noted that increased rainfall may have an impact on macroalgal production, such as kelp productivity, and may introduce a variety of nutrient loadings into nearshore ecosystems. In addition, invading short-lived algae species may be preferred over longer-lived kelp species that are typically thought to be ideal for cultivation due to variations in nutrient loading under varied precipitation.

Water stress from droughts can result in shortages and degradation, which can harm aquaculture productivity. The anticipated scarcity of water resulting from climate change is expected to escalate disputes over water among many user groups, including aquaculture, agriculture, households, and industries. This will have an impact on aquaculture sustainability in all its forms. Further research is necessary to determine how various fish species and life stages—particularly those with significant economic value—will react to variations in the pattern of precipitation.

2.5 Sea level rise

According to IPCC (2018) sea level rise forecasts, sea level increase under 1.5°C global warming will be around 0.1 m lower than under 2°C by 2100. Nonetheless, it is anticipated that this rise will continue after 2100, with the rate and amount of increase most likely to be determined by the future GHG pathways (IPCC, 2018). Sea level rise poses a threat to several coastal habitats, including salt marshes and mangroves, which are vital for preserving wild fish populations and providing seed for aquaculture. Programs for aquaculture breeding as well as the industry's ability to maintain itself economically would suffer as a result. Saline water intrusion from rising sea levels is expected to have an impact on aquaculture production infrastructure, including ponds, cages, tanks, and pens, especially in lowland areas. Freshwater fisheries, aquaculture, and agricultural productivity are all thought to be negatively impacted by groundwater. Salinization therefore makes aquaculture inappropriate for production in terms of the environment, resulting in greater production costs and lesser economic gains. Inland and marine aquaculture output may be threatened by changes in species composition, organism abundance and distribution, ecosystem productivity, and phenological shifts brought on by sea level rise. Additionally, aquaculture operations in coastal regions help society and the environment; nevertheless, the production and viability of the industry may be impacted by rising sea levels, both directly and indirectly. Positively, sea level rise may expand the regions that may support the brackish water culture of valuable species like mud crab and shrimp. This could contribute to the sustainability of aquaculture output by creating new options, especially for coastal locations.

2.6 Changes is sea surface salinity

According to scientists, salinity is viewed as a changeable parameter that reflects the entry of freshwater from precipitation, ice melting, river runoff, water loss through evaporation, and the mixing and circulation of ocean surface water with subsurface water. Changes in ocean circulation and rising temperatures can lead to higher evaporation, which in turn can cause variations in sea salinity. Alternatively, climate change might directly cause these variations. The capacity of the ocean to retain heat, as well as the transport of carbon and nutrients, may be impacted by these fluctuations in oceanic circulation and stratification. Aquaculture's environmental sustainability will be impacted since these fluctuations are anticipated to be brought on by climate change. According to researchers, the majority of aquatic creatures have a range of salinity levels at which they may thrive. Deviations from this range may result in mortality and decreased productivity. Salinity levels over the ideal range have been shown to diminish red blood cells, growth, and survival in stripped catfish, which may have an impact on the immune system of the fish. This means that variations in sea salinity are anticipated to have a detrimental impact on the profits realized by some aquaculture species, which may have an adverse effect on the social and economic dimensions of the sustainability of aquaculture production. However, in downstream coastal areas, the increased salinity effect has been closely connected with aquaculture production methods. Higher salinity has been shown to have detrimental consequences on freshwater prawn production performance. While scientists found greater susceptibility to bacterial invasion in oysters at reduced water salinity, observed higher mortality in immature clams. Additionally, it is noted that changes in salinity may have an impact on oysters' immune systems, specifically on hemocytes' (blood cells') capacity to fend off bacterial invasion from outside sources. Clams are said to be more tolerant of salt than other marine mollusks, although Baker *et al.*, (2013) pointed out that they cannot withstand extended exposure to either high or low salinity. Additionally, Scientists noted that changes in salinity may have an impact on oysters' immune systems, specifically on hemocytes' (blood cells') capacity to fend off bacterial invasion from outside sources. Clams are said to be more tolerant of salt than other marine mollusks, although Baker *et al.*, (2013) pointed out that they cannot withstand extended exposure to either high or low salinity. A changing climate due to the infiltration of saline water from marine environments has also been linked to increased mortality rates of abalone in farms, especially those that operate in coastal areas. Water salinity variations often result in greater mortality rates for a number of species, which might have an impact on the sector's social and economic sustainability through higher management costs and a rise in species losses.

2.7 Mitigation and adaptation options

According to current forecasts, there will be more dangers associated with climate change to human security, food security, livelihood, water supply, and economic growth. In order to reduce these risks and prepare for the changing environment, communities and business must seize new possibilities brought about by shifting resource availability. In order to develop resilience and cope with climate change as effectively and efficiently as possible, agricultural communities, ecosystems, and populations as a whole may benefit from mitigation and adaptation efforts. The goal of mitigation is to slow down or even stop the rate of climate change. This mostly entails cutting greenhouse gas emissions, with a concentration on carbon dioxide emissions, which make up over 60% of human-caused increases in emissions. In 2018, the IPCC stated that a mix of new and current technologies and practices, such as electrification, hydrogen, sustainable bio-based feedstocks, product substitution, and the use and storage of carbon capture, might reduce carbon dioxide emissions. By making the required changes to their production methods with the goal of decreasing GHG emissions, aquaculture farmers and other stakeholders may be able to significantly contribute to the mitigation of the consequences of climate change. To reduce air and water pollution, this specifically entails using ecologically friendly methods and technologies including solar energy, appropriate feeding techniques, and sustainable wastewater treatment. Building resilience to the outcomes and the ability to take advantage of new possibilities in a sustainable and morally responsible manner are the main goals of adaptation. However, given the complexities of exposure and vulnerability and their relationship to the socioeconomic and sustainable development of different sectors, there is no one-size-fits-all solution. The ability of producers in a particular country or region to adapt will determine whether or not adaptation in a changing climate is effective, according to the IPCC. For instance, because of their limited potential for adaptation, producers in underdeveloped nations have been expected to have more severe repercussions.

Conclusion

It's possible that the 21st century's global warming may lead to significant, possibly disastrous changes in the climate system that won't be felt until much later. A runaway greenhouse effect or a slowing or breakdown of the thermohaline circulation might be among these occurrences, leading to a cooling of the North Atlantic and the melting of the West Antarctic Ice Sheet in Western Europe and Eastern North America. It's unclear how much of a shift is needed to set off catastrophic consequences. The majority of the research that we looked at predicted that negative effects will worsen as the global mean temperature rises by about 3 to 4°C.

There is no consistent correlation found in the research between effects and global mean temperatures in the range of 0 to 3–4°C. It is evident that even modest temperature changes will have negative effects on coastal resources. Analysts' ongoing objective is to provide and take in data showing how better management of marine ecosystems and fisheries may surely contribute to adaptation to the effects of climate change. Comprehensive and clear information on the risks and uncertainties resulting from poor data quality and structural flaws in the assessment models must be included in management guidance. Managers and decision-makers are aware of adaptation measures, but political will and action are frequently absent. Fisheries and aquaculture management must embrace and abide by best practices, such as those outlined in the FAO Code of Conduct for Responsible Fisheries, in order to increase resilience to the effects of climate change and reap sustainable benefits. River basin, watershed, and coastal zone management require a stronger integration of these methods. To identify shifts and provide early notice of changes in the productivity of individual species as well as in the composition and operation of the ecosystems that support them, well-planned and trustworthy monitoring of fish stocks and the marine ecosystem is crucial.

The following ideas are put up for potential future fishery management

1. Adopt all-encompassing and integrated ecosystem management strategies for fisheries, aquaculture, catastrophe risk reduction, coast-sand seas, and climate change adaptation.
2. Switch to fuel- and environmentally-efficient methods for farming and fishing.

3. Get rid of subsidies that encourage excessive fishing and overfishing.
4. Carry out "local" vulnerability and risk assessments.
5. Connect and make aquaculture "climate-proof" with other industries.
6. Examine how aquatic habitats sequester carbon.
7. Address the opportunities and risks that the effects of climate change pose to the security of food and livelihood.

References

Baker, S., Hoover, E., and Sturmer, L. (2013). The Role of Salinity in Hard Clam Aquaculture. Gainesville, FL: University of Florida.

Blanchard, J. L., Watson, R. A., Fulton, E. A., Cottrell, R. S., Nash, K. L., Bryndum Buchholz, A., (2017). Linked sustainability challenges and trade-offs among fisheries, aquaculture and agriculture. Nat. Ecol. Evol. 1, 1240–1249. doi: 10.1038/s41559-017-0258-8

Cochrane, K., De Young, C., Soto, D. T., and Bahri, D. T. (2009). Climate Change Implications for Fisheries and Aquaculture: Overview of Current Scientific Knowledge. FAO Fisheries and Aquaculture Technical Paper No. 530. Rome: FAO.

IPCC (2013). Summary for Policymakers, The Physical Science Basis. Contribution of Working Group I to the Fifth Assessment Report of the Intergovernmental Panel on Climate Change. Cambridge; New York, NY: Cambridge University Press.

IPCC (2014). Climate change 2014: Synthesis Report. Contribution of Working Groups I, II and III to the Fifth Assessment Report on the Intergovernmental Panel on Climate Change. Core writing team, R. K. Pachauri and L.A. Meyer. Geneva: Intergovernmental Panel on Climate Change, 151 pp. Available online at: http://www.ipcc.ch/pdf/assessmentreport/ar5/syr/SYR_ AR5_FINAL_full_wcover.pdf

IPCC (2018). Global Warming of 1.5◦C. An IPCC Special Report on the Impacts of Global Warming of 1.5◦C Above Pre-Industrial Levels and Related Global Greenhouse Gas Emission Pathways, in the Context of Strengthening the Global Response to the Threat of Climate Change, Sustainable Development, and Efforts to Eradicate Poverty, eds V.

IPCC (2019). Special Report on Climate Change, Desertification, Land Degradation, Sustainable Land Management, Food Security, and Greenhouse Gas Fluxes in Terrestrial Ecosystems, Summary for Policymakers Approved Draft. Geneva: IPCC.

Ngoan, L. D. (2018). Effects of climate change in aquaculture: case study in Thua Thien Hue Province, Vietnam. Biomed. J. Sci. Tech. Res. 10:2018. doi: 10.26717/BJSTR.2018.10.001892

Rutkayova, J., Vácha, F., Maršálek, M., Beneš, K., Civišová, H., Horká, P., et al (2017). Fish stock losses due to extreme floods–findings from pond-based aquaculture in the Czech Republic. J. Flood Risk Manage. 11, 351–359. doi: 10.1111/jfr3.12332

Sae-Lim, P., Kause, A., H. Mulder, A., and Olesen, I. (2017). Breeding and genetics symposium: climate change and selective breeding in aquaculture. J. Anim. Sci. 95, 1801–1812. doi: 10.2527/jas2016.1066

5

Aquaculture and Its Impact on Environment

Shruti Samson and Kamin Alexander

Department of Biological Sciences, Sam Higginbottom University of Agriculture Technology and Sciences, Prayagraj-211007, Uttar Pradesh, India

Abstract

Aquaculture, the cultivation of aquatic organisms such as fish, shellfish, and seaweed, has grown exponentially in recent decades due to overfishing and the increasing demand for seafood. This chapter provides a comprehensive overview of aquaculture practices, including its methods, species, and technological advancements. It also delves into the environmental impacts associated with aquaculture, both positive and negative. Key issues such as habitat destruction, water pollution, genetic diversity, and using wild fish for feed are discussed, along with mitigation strategies and sustainable practices. The chapter concludes with a discussion on policy frameworks and future directions for minimizing the environmental footprint of aquaculture while maximizing its potential to contribute to global food security.

1. Introduction

Aquaculture, or the farming of aquatic organisms, has become a significant component of the global food system. It encompasses the breeding, rearing, and harvesting of fish, shellfish, algae, and other aquatic organisms in various aquatic environments including freshwater, brackish water, and marine water. Aquaculture's rapid expansion has been driven by the decline in wild fish stocks due to overfishing, as well as the growing global population and increased per capita consumption of seafood (FAO, 2020).

In recent years, aquaculture has outpaced traditional capture fisheries in terms of growth, making it one of the fastest-growing food production sectors worldwide. This growth is largely attributed to technological advancements and innovations in breeding, feeding, and disease management, which have significantly improved productivity and sustainability. The FAO (2020) reports that aquaculture now supplies over 50% of the global seafood consumed, highlighting its crucial role in meeting the dietary needs of billions of people.

Despite its growth, aquaculture is not without its challenges. Environmental concerns, such as habitat destruction, water pollution, and the impact on wild fish populations, have sparked significant debate. Additionally, socioeconomic issues, including the rights of local communities and labor conditions, require careful consideration. Addressing these challenges through sustainable practices and robust regulatory frameworks is essential to ensure that aquaculture can continue to expand without compromising the health of our ecosystems and societies. This chapter aims to provide a detailed exploration of aquaculture practices and their environmental impacts, as well as potential solutions to enhance sustainability in the industry.

2. Aquaculture Practices

Aquaculture involves a variety of practices that differ based on the environment and species being cultivated. These practices have evolved significantly over time, incorporating advancements in technology and sustainability. Understanding the types of aqua-culture is essential for appreciating the diversity and complexity of the industry.

3. Types of Aquaculture

Aquaculture can be broadly classified into several types based on the environment in which it is practiced and the species that are cultivated. Each type has its unique characteristics, advantages, and challenges.

3.1 Mariculture

Fig. 1. Aerial view of a mariculture farm in a bay. (The Pearl Protectors, 2023)

3.2 Freshwater Aquaculture

Freshwater aquaculture involves farming species that live in freshwater environments such as ponds, rivers, and lakes. This type of aquaculture is one of the most traditional and widespread forms of fish farming. Species commonly farmed in freshwater systems include tilapia, catfish, and carp. Freshwater aquaculture can be practiced in various systems, ranging from small-scale backyard ponds to large, commercial operations. It is particularly important in inland areas where access to marine resources is limited. Freshwater aquaculture provides significant economic and nutritional benefits, especially in developing countries. However, it also faces challenges such as water use conflicts, pollution from waste discharge, and the risk of spreading diseases to wild populations (Tacon & Metian, 2015).

4. Integrated Multi-Trophic Aquaculture (IMTA)

Integrated Multi-Trophic Aquaculture (IMTA) is an innovative approach that involves cultivating multiple species from different trophic levels in the same system. In IMTA systems, the byproducts (wastes) from one species serve as inputs (fertilizers, food) for another species. For example, fish waste can be utilized by filter-feeding shellfish and seaweed, creating a more balanced and sustainable ecosystem. This method mimics natural ecosystems where different species coexist and interact, promoting resource efficiency and reducing environmental impacts. IMTA can enhance productivity and sustainability by improving water quality, reducing feed costs, and increasing biodiversity. It represents a significant advancement in aquaculture practices, aiming to create a more circular and environmentally friendly system (Naylor *et al.*, 2009).

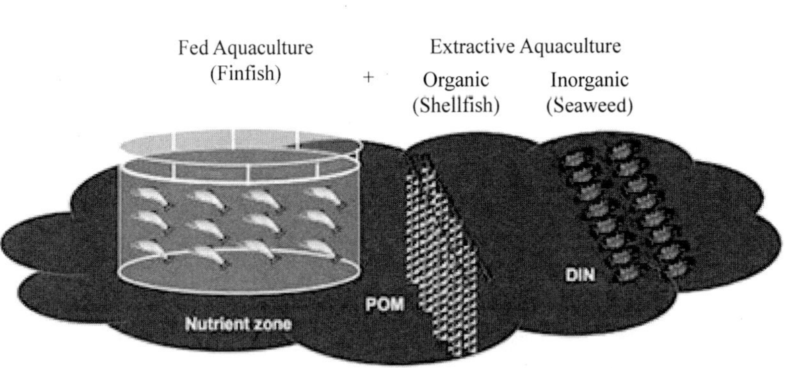

Fig. 2. Conceptual diagram of Integrated Multi-Trophic Aquaculture (IMTA) (Soto, 2009)

5. Techniques and Technologies

Advancements in techniques and technologies have played a crucial role in enhancing the efficiency and sustainability of aquaculture. These innovations have allowed the industry to meet growing demand while minimizing environmental impacts.

5.1 Pond Systems

Pond systems are one of the oldest and most widely used methods of aquaculture. In this system, species are cultivated in natural or man-made ponds. Ponds can vary in size and depth, and they are typically managed to optimize water quality and feed efficiency. Pond aquaculture is suitable for a variety of freshwater species such as carp, tilapia, and catfish. This method is advantageous because it is relatively simple and cost-effective, especially for small-scale farmers. However, it can also lead to environmental challenges such as water pollution from nutrient runoff and the potential spread of diseases if not properly managed (Boyd & McNevin, 2015).

5.2 Recirculating Aquaculture Systems (RAS)

Recirculating Aquaculture Systems (RAS) are closed-loop systems that recycle water within the system, significantly reducing water use and environmental discharge. In RAS, water is continuously filtered and treated to remove waste products before being recirculated back to the fish tanks. This method allows for greater control over water quality and environmental conditions, making it possible to farm species intensively in areas where water resources are limited. RAS is highly efficient and sustainable, but it requires significant investment in frastructure and technology. It is particularly suited for high-value species such as salmon, trout, and shrimp (Boyd & McNevin, 2015).

6. Open Net Pens/Cages

Open net pens or cages are used in marine and freshwater environments, where fish are reared in enclosures within their natural habitat. This method allows for the cultivation of species such as salmon, trout, and seabass in coastal and offshore areas. Open net pens are advantageous because they take advantage of natural water exchange, which can help to dilute waste and provide a constant supply of fresh water. However, this method also poses significant environmental risks, including pollution from uneaten feed and fish waste, the potential for fish escapes, and the spread of diseases and parasites to wild populations.

Fig. 3. Three circular fish cages float in a lake (Soto, 2009)

7. Flow-Through Systems

Flow-through systems involve water flowing through the culture units once before being discharged. These systems require access to high-quality water sources and are typically used for species that require specific environmental conditions, such as trout and salmon. Flow- through systems are advantageous because they provide a continuous supply of fresh water, which can improve fish health and growth rates. However, they also pose environmental challenges, particularly in terms of water use and the discharge of nutrient-rich effluents into surrounding water bodies (Tacon & Metian, 2015).

These various techniques and technologies illustrate the diversity and complexity of aquaculture practices, each with its own set of advantages and challenges. Understanding these practices is essential for developing strategies to enhance the sustainability and efficiency of the aquaculture industry.

8. Environmental Impacts of Aquaculture

While aquaculture presents a solution to overfishing and supports food security, it also poses environmental challenges. These impacts can be both negative and positive.

8.1 Negative Impacts

The rapid expansion of aquaculture has led to several environmental concerns that need to be addressed to ensure sustainable practices.

Habitat Destruction: The construction of aquaculture facilities can lead to the destruction of critical habitats like mangroves and wetlands, which are essential for biodiversity and coastal protection.

Water Pollution: Nutrient runoff from uneaten feed and fish waste can lead to eutrophication, harmful algal blooms, and hypoxia in surrounding water bodies (Tacon & Metian, 2015).

Genetic Pollution: Escapees from aquaculture operations can interbreed with wild populations, potentially diluting genetic diversity and leading to the spread of diseases.

Use of Wild Fish for Feed: Many farmed fish, particularly carnivorous species, are fed fishmeal and fish oil derived from wild-caught fish, putting additional pressure on wild fish stocks (Naylor *et al.*, 2009).

Disease and Parasites: High stocking densities can facilitate the spread of diseases and parasites, which can also impact wild populations.

8.2 Positive Impacts

Despite its challenges, aquaculture also offers several environmental benefits that can be harnessed through sustainable practices.

Reduction of Fishing Pressure: Aquaculture can reduce pressure on overfished wild stocks by providing alternative sources of seafood.

Economic Benefits: It can provide livelihoods and economic benefits to coastal and rural communities.

Innovative Practices: Methods like IMTA and RAS can mitigate environmental impacts and promote sustainable production (Boyd & McNevin, 2015).

Sustainable Aquaculture Practices: To minimize the environmental impacts of aquaculture, various sustainable practices have been developed and implemented. These practices aim to balance production with ecological health.

9. Best Management Practices (BMPs)

Best Management Practices (BMPs) are essential for promoting sustainability in aquaculture. They encompass a range of strategies and techniques designed to reduce environmental impacts.

Site Selection: Choosing appropriate sites to minimize habitat destruction and water pollution.

Feed Management: Using sustainably sourced feed and optimizing feed conversion ratios to reduce waste.

Disease Management: Implementing biosecurity measures and using vaccines to prevent disease outbreaks.

Waste Management: Techniques to manage and utilize waste products, such as using fish waste as fertilizer.

Certification and Standards

Certification programs and standards play a vital role in promoting sustainable aquaculture by setting benchmarks for environmental and social responsibility. The Aquaculture Stewardship Council (ASC) is one such organization that provides certification programs aimed at ensuring environmentally and socially responsible aquaculture practices (ASC, 2023). Similarly, the Global Aquaculture Alliance (GAA) promotes sustainable aquaculture through its Best Aquaculture Practices (BAP) certification, which sets rigorous standards for environmental sustainability, food safety, social responsibility, and animal welfare (GAA, 2023). These certification programs help to drive improvements in aquaculture operations, encouraging producers to adopt practices that minimize environmental impact and enhance the social and economic benefits of aquaculture.

References

Aquaculture Stewardship Council. (2023). ASC Standards. https://www.asc-aqua.org/

Boyd, C. E., & McNevin, A. A. (2015). Aquaculture, Resource Use, and the Environment. Wiley-Blackwell.

Food and Agriculture Organization of the United Nations. (2020). The State of World Fisheries and Aquaculture 2020. FAO. https://www.fao.org/publications/sofia/2020/en/

Global Aquaculture Alliance. (2023). Best Aquaculture Practices. https://www.aquaculture-alliance.org/

Naylor, R. L., Hardy & Nichols, P. D. (2009). Feeding aquaculture in an era of finite resources. Proceedings of the National Academy of Sciences, 106 (36), 15103-15110. https://doi.org/10.1073/pnas.0905235106

Soto, D. (2009). Integrated Mariculture. Food & Agriculture Org.

Tacon, A. G. J., & Metian, M. (2015). Feed matters: Satisfying the feed demand of aquaculture. Reviews in Fisheries Science & Aquaculture, 23(1), 1-10. https://doi.org/10.1080/23308249.2015.1007003

The Pearl Protectors. (2023). mariculture. https://pearlprotectors.org/introduction-to- mari-culture/.

6

Bioprospecting in Fisheries and Aquaculture

Ridhdhisa R. Barad[1*], Ajay R. Ram[1], S.I. Yusufzai[1], Foram Vala[1], Krina D. Patel[1], Vishal K. Solanki[1] and Durgesh Kumar Verma[2]

[1]*Department of Aquaculture, College of Fisheries Science, Kamdhenu University Veraval-362265, Gujarat, India*

[2]*ICAR-CIFRI, Regional Centre, Prayagraj-211002, Uttar Pradesh, India*

Abstract

The systematic exploration of the biological diversity of valuable compounds and organisms has emerged as a promising avenue for aquaculture and fisheries. As global demand for seafood continues to rise, coupled with the increasing challenges of overfishing and environmental degradation, the quest for sustainable solutions has intensified. Bioprospecting in aquaculture involves the identification and utilization of novel marine organisms, including microorganisms, algae, and invertebrates, for the development of aquafeeds, pharmaceuticals, and bioactive compounds. The exploration of untapped genetic resources in fisheries can lead to the discovery of resilient and economically valuable fish species, contributing to the diversification of aquaculture production. The integration of bioprospecting into aquaculture and fishery practices holds immense promise for fostering sustainable development, ensuring food security, and mitigating the environmental impacts of traditional practices. However, successful implementation requires a balanced approach that considers economic, social, and environmental factors, ultimately guiding the industry toward a more resilient and ecologically responsible future.

Keywords: *Aquaculture, Marine Bioprospecting, Cryoprotectants, Nano-materials, AFPs*

1. Introduction

What is Bioprospecting?

Biodiversity prospecting, alternatively referred to as bioprospecting, involves the exploration and development of new products sourced from the natural environment. Scientists actively seek natural substances from plants, animals, microorganisms, and other sources that could have beneficial applications for humans. The primary objective is to leverage these discoveries to improve society through commercialization and the creation of various goods. These valuable resources have applications in industries such as cosmetics, agriculture, and pharmaceuticals. Although bioprospecting can take place in any biodiverse region, it is most commonly pursued in areas with exceptionally rich biodiversity. Regions with diverse ecosystems are more likely to harbor organisms with properties suitable for bioprospecting (Rollo & Cheprasov, 2023).

1.1 Why is it needed?

Finding new, naturally occurring resources and products that humans can use is a fundamental goal of bioprospecting. Enhancing human well-being via medication and improved diet is an important focus area. Given the scarcity of known or current chemicals for the creation of pharmaceuticals for human use, bioprospecting is a key method for determining the leads for drug development. Originality and complexity that can be altered in a laboratory can be found in nature.

Bioprospecting in aquaculture and fisheries involves the exploration and utilization of biological resources from aquatic environments to enhance and sustainably manage activities related to fishery and aquaculture industries. This multifaceted approach aims to discover, develop, and apply valuable genetic, biochemical, and ecological components derived from aquatic organisms including fish, mollusks, and other species. Scientists and researchers engage in bioprospecting within these domains to uncover novel solutions for improving aquaculture practices, ensuring food security, and addressing challenges in fishery management.

The diverse ecosystems of oceans, rivers, lakes, and other aquatic environments offer rich sources of untapped biological potential. Bioprospecting in aquaculture seeks to identify and harness the unique traits of aquatic organisms that can contribute to the development of new breeds, improve disease resistance, and enhance the nutritional profiles of farmed fish. Additionally, it explored the potential of bioactive compounds from aquatic species for applications in pharmaceuticals, nutraceuticals, and other industries.

This review highlights the importance of bioprospecting in aquaculture and fisheries, emphasizing its role in advancing sustainable practices, promoting biodiversity conservation, and addressing emerging challenges in the rapidly evolving field of aquatic resource management.

1.1.1 There are three aspects of bioprospecting (Bhooleshwari 2023)

1. Chemical prospecting: Pesticides, medications and pharmaceuticals, food additives, nutraceuticals, cosmetics, and skincare.
2. Gene prospecting: crop development, fermentation, genetic engineering, and cell culture.
3. Bionic prospecting: Bioengineering, Bio-modeling, Sensor technologies, Designs, and Architecture.

1.1.2 There are three main steps to consider in the process of bioprospecting

1.1.2.1 Collection: Gathering of natural materials was the first stage. Scientists usually gather samples from natural resources. To do this, they must travel to an area where resources are typically found and collected. They occasionally gather data on the application of resources in a certain field. Indigenous knowledge is information derived from local people. Scientists have frequently discovered resources that locals have been utilizing for a long time.

1.1.2.2 Scientific Analysis and Research: Following collection, samples were examined to see what applications they might have and how they might be made profitable. Scientists will examine and investigate the chemical makeup of the material to determine any possible advantages and the most effective way to obtain them. Additionally, they want to guarantee that nothing in it will be dangerous for humans to utilize.

1.1.2.3 Commercialization: For the gathered resources to be useful to individuals, they must be made available commercially. The term "commercialization" refers to the process of managing something so that it is produced to generate revenue. To make one of the resources easily accessible for purchase, it may be commercialized and offered for sale at Walmart.

2. Bioprospecting process

2.1 Phase 1

- Sample collection, either at the site or through an agency, marine, terrestrial, or both

2.2 Phase 2

- Compound identification;
- Compound isolation;
- Characterization and synthesis of certain chemicals

2.3 Phase 3

- Bioactivity screening
- Bioactivity Confirmation

2.4 Phase 4

- Development of product and testing
- Commercialization

3. Potential Benefits to Bioprospecting in Aquaculture

Modern bioprospecting, which entails seeking economically valuable genetic and biochemical resources within biodiversity, not only helps preserve natural landscapes and wildlife by supporting conservation efforts but also continues our age-old tradition of investigating nature to enhance human well-being. Additionally, bioprospecting should prioritize advancing conservation and economic goals that are crucial for driving agricultural and medical innovations needed to address diseases and sustain a growing global population (Cox and Balick, 1994).

For many years, natural products, such as plants, animals, microbes, and marine species have become significant sources of prospective therapeutic leads. Three primary search strategies–rational, chemo-rational, and random—have been used to isolate several bioactive compounds from these sources. For instance, a random screening strategy led to the discovery of conocurvone, an anti-HIV drug. A bio-rational approach has been used to identify drugs such as morphine, quinine, and artemisinin (Rishton, 2008).

Exploring new fish or shellfish species presents an opportunity to discover advantageous characteristics such as rapid growth, disease resistance, and adaptability to specific environmental conditions, which could significantly benefit commercial aquaculture. Marine life, in particular, may yield bioactive compounds with therapeutic qualities, such as antimicrobials, antivirals, or antioxidants, that have potential health benefits for both humans and animals.

Aquatic plants and algae may also contain valuable elements such as pigments or oils, with applications in various industries, including food and cosmetics. Nevertheless, a sustainable approach is crucial for bioprospecting. This entails evaluating potential environmental consequences and ensuring that any advantages gained from commercialization are distributed fairly and justly among the communities and nations from which the organisms originate (Akiwumi, 2022). Adhering to international agreements and national laws related to biodiversity and intellectual property is essential for conducting bioprospecting activities.

Bioprospecting in aquaculture involves systematic search, collection, and analysis of biological resources from aquatic environments to identify valuable traits or compounds that can be applied to enhance and sustain aquaculture practices. This method encompasses a diverse array of aquatic life forms, such as fish, mollusks, and other marine species. The primary objectives of bioprospecting in aquaculture include improving the productivity, disease resistance, and environmental adaptability of cultured species as well as exploring bioactive compounds with potential applications in various industries.

3.1 Aquaculture and Plant-Derived Compounds

The regulated production of water-dwelling organisms, such as fish, shellfish, and aquatic plants, is known as aquaculture. Compounds originating from plants can be employed in aquaculture in various ways that are advantageous to the environment and farmed creatures. These are a few instances

3.1.1 Fish feed: To provide farmed fish with the vital nutrients they need, plant-derived substances are frequently employed as ingredients. Corn, wheat, pea protein, and soybean meal are common plant-based components. These components can be prepared and processed to satisfy the unique dietary requirements of the various fish species.

3.1.2 Phytochemicals: Carotenoids, polyphenols, and phycobilins are only a few of the many compounds found in aquatic plants, such as algae and seaweeds. Fish health and immune system function can be enhanced by using phytochemicals produced from aquatic plants as functional additives or natural pigments.

3.1.3 Natural medicines include substances with antibacterial, antifungal, and antiparasitic qualities obtained from plants. Certain plant extracts or essential oils can be utilized in aquaculture as natural substitutes for artificial antibiotics or disease-causing chemicals, for both treatment and prevention.

3.1.4 Water Treatment: It is well known that Aquatic plants can filter water and absorb nutrients. They may be utilized in aquaculture systems to assist in managing algal blooms and enhancing water quality by lowering surplus nutrients, such as phosphorus and nitrogen. In addition to releasing oxygen and absorbing carbon dioxide, aquatic plants also help oxygenate water.

3.1.5 Soil Nutrient Cycling: Integrated multitrophic aquaculture (IMTA) and aquaponics are two plant-based farming systems that may be combined with aquaculture. In these systems, fish farming effluent rich in nutrients is used by plants as a fertilizer. By drawing nutrients from the water that would otherwise build up in the plants, the environmental effect of aquaculture is lessened, and a closed-loop system is established. Aquaculture operations can be improved in terms of sustainability, environmental consequences, and the health and well-being of farmed aquatic species by introducing plant-derived substances.

3.2 Bioprospecting of microalgae for Aquaculture feed

Microalgae are simple, primitive, autotrophic, bacterial or eukaryotic, aquatic, or subaerial primary producers that live in parasitic, symbiotic, or free-living environments. They have almost entirely taken over biotech markets in the domains of nourishment, fuel, feed, environmental services, etc. The high nutrient content and simple digestion of microalgae make them an ideal source of food and feed. Feed is the most significant exogenous element influencing animal growth and aquaculture costs. The cost and carbon footprint of aquafeed production can be decreased by substituting fish-derived materials with less expensive and environmentally friendly raw resources, such as microalgae. This would enhance the advantages and sustainability of the aquaculture sector. Algal production in animal feed constitutes over 30% of global algal output. Research into nutrition and safety has demonstrated the potential of algal biomass as a sustainable feed supplement. Among the most frequently used microalgae in aquaculture are varieties such as Chlorella, Isochrysis, Tetraselmis, Pavlova, Chaetoceros, Phaeodactylum, Nannochloropsis, Thalassiosira, and Skeletonema. Apart from their growth-promoting properties, algal supplements in aquafeeds also improve fish and shrimp pigmentation, health, and performance (Achankunju, 2016).

4. Scope of Bioprospecting

Marine bioprospecting contracts can be a useful tool for recognizing traditional knowledge-based research and other activities as well as for developing strong, profitable processes. It is also depicted that contractual frameworks, in conjunction with the customary knowledge of coastal communities, can ensure better and more sustainable use of marine resources. The demand for bio-derived products, including pharmaceuticals, chemicals, and renewable fuels, that support eco-friendly technology and sustainable progress is increasing alongside global population growth. In turn, this need would boost the efforts related to marine bioprospecting. As a result, creating contracts and processes as a fundamental part of the marine bioprospecting industry is easy and will spark a variety of ideas and interests among entrepreneurs and stakeholders (Sylvia, 2022).

The scope of bioprospecting in aquaculture and fisheries is broad and holds significant potential for advancing sustainable practices, improving productivity, and addressing challenges in these industries. Key aspects that highlight the scope of bioprospecting in aquaculture and fisheries are as follows.

4.1 Genetic Improvement

Bioprospecting allows the identification and selection of species with desirable genetic traits, such as disease resistance, fast growth, and adaptability to varying environmental conditions. This information could be used to enhance selective breeding programs, resulting in improved and more resilient aquaculture stocks.

4.2 Disease Management

The search for bioactive compounds with antimicrobial, antiviral, and antioxidant properties in aquatic organisms can contribute to the development of disease management strategies in aquaculture. These compounds may provide natural alternatives to conventional treatments, thereby promoting healthier and more sustainable aquaculture practices.

4.3 Nutritional Advancements

Aquatic plants, algae, and other organisms identified through bioprospecting may contain valuable nutritional compounds such as pigments and oils. These can be incorporated into fish feed to enhance the nutritional content of farmed fish, thereby contributing to the development of more balanced and nutritious aquaculture products.

4.4 Environmental Adaptability

Bioprospecting can help to identify species that demonstrate resilience to specific environmental conditions, such as temperature variations or water quality parameters. This information can be valuable in promoting aquaculture practices that are better adapted to local environments, thereby reducing the need for external interventions.

4.5 Sustainable Practices

Bioprospecting supports the development of sustainable aquaculture practices by exploring the biodiversity of the aquatic ecosystems. This includes responsible resource management, minimization of environmental impacts, and ensuring the long-term viability of aquaculture operations.

4.6 Economic Opportunities

Bioprospecting presents economic opportunities through the discovery and commercialization of novel products. Compounds with applications in pharmaceutical, cosmetic, and other industries can create new revenue streams for aquaculture and fisheries, potentially benefiting local economies.

4.7 Conservation of Biodiversity

The careful and sustainable exploration of biodiversity in aquatic environments has promoted conservation efforts. Bioprospecting activities can contribute to an improved understanding of ecosystems, and the importance of preserving biodiversity for the health and sustainability of aquaculture and fisheries (Harinkhede *et al.*, 2023).

4.8 Legal and Ethical Considerations

Bioprospecting requires adherence to international agreements and national laws on biodiversity and intellectual property. Establishing legal and ethical frameworks ensures that the exploration and utilization of genetic resources are conducted responsibly and with respect to local communities.

5. The Nature And Scope of Marine Bioprospecting

Marine bioprospecting involves the investigation of biological materials in the oceanic environment to identify commercially valuable genetic and biochemical characteristics. This field is expanding rapidly in both research and industry. Due to the extensive biodiversity and uniqueness of marine ecosystems, there is substantial economic promise. Additionally, these environments harbor potentially active chemical compounds that could lead to the development of distinctive natural products (Davies and Sunassee, 2012; Leal *et al.*, 2012; Martins *et al.*, 2014).

In the 1970s and the 1980s, marine macroalgae received primary attention in research on marine natural products, but since then, microorganisms and marine invertebrates have become focal points of bioprospecting endeavors (Leal *et al.*, 2013). There is an increasing interest in the potential of marine microalgae, especially in the food and cosmetics sectors, as evidenced by the growing attention in research studies (Martins *et al.*, 2014). Numerous marine organisms produce natural substances as a form of chemical defence against predators or as a response to competition among species for limited resources (Davies and Sunasee, 2012).

6. Research, Development and Commercial Application

In general, investments in marine bioprospecting are both costly and risky, especially when conducted in deep-water environments where success rates are low and significant regulatory hurdles exist for product approval. For example, it took nearly three decades for Prialt®, the initial medication sourced from a marine organism, the venom of a tropical marine cone snail used to paralyze its prey, to obtain approval in the US as a treatment for chronic pain (Marris 2006, Molinsky *et al.*, 2009).

Microalgae are extensively sought for the mass production of polyunsaturated fatty acids, which are utilized in nutritional supplements. Microalgae are utilized in various biomedical applications, including the production of polysaccharides for use in the food and health sectors, as well as in the manufacturing of fluorescent proteins, bone fillers, and bioceramic coatings. Furthermore, "extremozymes" derived from microalgae have applications in various industrial processes (Arrieta *et al.*, 2010; Silva *et al.*, 2012; Laird, 2013).

The research and development of products face considerable hurdles in securing a sustainable marine source, often found in sparsely populated and remote growth regions. While the initial bioprospecting phases typically involve collecting small biomass samples, the retrieval of noteworthy species or scaling up for commercial purposes may pose significant ecological risks. The conservation status of many species and the potential damage to their ecosystems are primary areas of concern (Costello *et al.*, 2010; Global Ocean Commission 2013).

6.1 Marine Bioprospecting

1. The practice of bioprospecting in the terrestrial environment dates back hundreds of years when samples of the aquatic environment were collected and screened for commercial use.
2. The study of the great chemical and biological diversity present in marine species that live in the oceans is known as marine bioprospecting, sometimes referred to as marine natural product research.
3. A wide range of commercial items obtained from marine organisms has been introduced into the market, such as those listed below (Verma *et al.*, 2009; Verma *et al.*, 2023).

6.2 Marine Products

6.2.1 Cod liver oil

Cod liver oil, sourced from species such as Gadus callarius and Gadus morrhua, contains vitamins A and D and omega-3 fatty acids. It is utilized for treating various conditions, such as glaucoma, hypertension, osteoarthritis, kidney disease, and wound healing, as well as for lowering blood lipid levels. Its consumption has been shown to enhance nutrition and bone mineralization in patients with tuberculosis and rickets because of its rich vitamin A and D content and easily digestible fats. Cod liver oil can also be topically applied to burns and wounds, and serves as a dietary supplement for children. A significant historical discovery highlighted the pivotal role of cod liver oil in understanding and combating rickets, with research in 1922 establishing its protective and therapeutic effects alongside exposure to sunlight in preventing disease in infants.

Therefore, using a nutrigenomic method, consuming fish and fish oil may promote gene expression to maintain brain function at one age. According to Rajakumar (2003), it is offered as E-COD Omega-COD Plus.

6.2.2 Shark cartilage

Derived from the tough material found in shark skeletons, specifically cartilage. It is sold under several brand names, such as Tiburon and Cartacin, and is used to prevent several illnesses including cancer (Molinsky, 2009).

6.2.2.1 Shark cartilage is used to treat cancer, especially Kaposi's sarcoma, a disease that is more prevalent in HIV-positive individuals. Additionally, shark cartilage has been utilized to treat wound healing, psoriasis, arthritis, diabetic retinal damage, and intestinal inflammation (enteritis). Some people treat psoriasis and arthritis by directly applying shark cartilage to their skin (Ostrander, 2004).

6.2.2.2 Cancer: Most studies indicate that oral shark cartilage is not beneficial for patients with advanced non-Hodgkin's lymphoma or previously addressed tumors of the brain, colon, breast, lung, prostate, or other organs. There have been reports that shark cartilage may reduce Kaposi sarcoma tumors (Davies *et al.*, 2012).

6.2.2.3 Osteoarthritis: Products that combine chondroitin sulfate, glucosamine sulfate, and camphor with shark cartilage have been claimed to have beneficial effects when topically administered to lessen the symptoms of arthritis. Any symptom improvement, however, is probably the result of camphor's action rather than other substances. Furthermore, no study has demonstrated that the skin absorbs shark cartilage.

6.2.2.4 Advanced kidney cancer: Renal cell carcinoma, a type of kidney cancer, can be treated with shark cartilage extract. This product has been granted "Orphan Drug status" by the FDA for the treatment of renal cell carcinoma. The Orphan Drug Law provides pharmaceutical companies with the extra motivation to research medications for uncommon diseases.

6.2.2.5 Psoriasis: Emerging research indicates that an extract from shark cartilage, applied topically or consumed orally, may alleviate the appearance and reduce itching in plaque psoriasis (Arnaud□Haond *et al.*, 2011).

6.2.3 Fish-Based Protein

Fish proteins are highly digestible and exhibit biological and growth-promoting properties. Consequently, human nourishment is crucial. The amino acid makeup of this protein is more consistent than that of other animal-derived proteins. Fish protein is rich in amino acids, such as lysine and methionine, potentially surpassing chicken protein and showing slight superiority over egg albumen, bean protein, and casein. Fish muscle is a great source of nutrient-dense, readily digested proteins, and has a great amino acid makeup.

However, using fish as a primary raw food material creates special challenges for food processing owing to its severe perishability and variable chemical composition. The primary source of protein, comprising 15–25 percent, is extracted from the muscle tissue of fish. (Ryan *et al.*, 2011)

6.2.3.1 Fish meal

Fish meals can be created from a wide variety of seafood, but they are commonly derived from small marine fish that are caught in the wild and possess high oil and bone content, making them unsuitable for direct human consumption.

The phrase "industrial" refers exclusively to fish that is consumed for fish meal. Additional sources of fishmeal include bycatch and byproducts of trimming created during the processing of different seafood items intended for direct human consumption (also known as fish waste or offal). Fishmeal serves as a valuable protein source for poultry and is incorporated into the feed for pigs and poultry. It contains high levels of essential amino acids such as lysine and methionine, along with a beneficial combination of unsaturated fats, specific minerals, and vitamins. (Rosendal 2006).

6.2.3.2 Marine macroalgae serves as a reservoir for bioactive compounds and nutraceuticals

Over the past decade, there has been a notable increase in interest regarding bioactive compounds sourced from algae and their application as functional food ingredients, as evident in scientific literature and patent filings. These advancements have demonstrated the potential of bioactive compounds extracted from seaweeds to promote and improve human health and well-being. The pharmaceutical and healthcare sectors find various nutraceuticals, functional food supplements, and biomedical products derived from marine algae because of their numerous health benefits and potential to address various life-threatening illnesses. It has been found that certain components of marine algae act as chemical mediators that modulate inflammation and regulate stimuli associated with mammalian defense mechanisms.

Brown and red algae, classified as Phaeophyceae and Rhodophyceae, respectively, have been recognized as potential reservoirs of bioactive compounds among the numerous marine algal species worldwide. Although the global multi-billion-dollar industry revolves around marine algae, their bioactive potential remains largely unexplored. Historically, the therapeutic benefits of marine algae have been limited to traditional and traditional remedies for millennia. However, in recent years, industries, such as cosmetics, pharmaceuticals, and food, have increasingly focused on the discovery and synthesis of chemicals derived from marine algae.

Products of marine origins that are available in the markets for food, medicine, and personal care *(Source: Global Ocean Commission, 2013)*

Product	**Organism**	**Category**
Brominated furanone (quorum sensing inhibitor)	Delisea pulchra (red alga)	Biofilm inhibitor
Carotenoids (anti-oxidant)	Dunaliella salina (microalga)	Nutrition
ω-3 fatty acids	Crypthecodinium cohnii (microalga)	Nutrition
Mycosporine-like amino acids (UV absorbing)	Coral zooxanthellae	Sunscreen
Pseudopterosins (anti-inflammatory)	Pseudopterogorgia elisabethae (soft coral)	Cosmetic
Venuceane (anti-free radicals)	Thermus thermophilus (bacterium)	Cosmetic
Halaven® (cancer)	Halichondria okadai (sponge)	Therapeutic
Prialt® (neuropathic Pain)	Conus magus (mollusc)	Therapeutic
Plinabulin (cancer)	Aspergillus sp. (fungus)	Therapeutic
Salinisporamide (cancer)	Salinispora tropica (bacterium)	Therapeutic
Yondelis® (cancer)	Ecteinascidia turbinate (ascidian)	Therapeutic

7. Nanomaterials: Reforming Bioprospecting To Support Enduring Drugs

The methodical search for new biological resources with potential use in biotechnology, agriculture, medicine, and other fields has long been a pillar of scientific inquiry and advancement. This process is known as bio-prospecting. It involves the identification and long-term use of a variety of biological substances, from genetic sequences to bioactive chemicals sourced from natural sources. Bioprospecting has undergone a paradigm shift in recent years because of the incorporation of nanomaterials, which has transformed its techniques and increased its potential. Our investigation in this study was based on this change. Combining nanomaterials with bioprospecting is a novel way to improve many parts of the bioprospecting process by leveraging the special qualities of materials at the nanoscale, which are generally smaller than 100 nm in at least one dimension. Nanomaterials, including nanoparticles, nanocomposites, and nanoscale probes, have the potential to completely change the gathering, extraction, examination, and use of biological resources from various ecosystems (*et al.*, 2019).

7.1 Cryoprotectants

Bioprospecting in harsh settings demonstrates a variety of creatures and ecosystems. Numerous plants, animals, and invertebrate species from high-altitude or latitude environments have been shown to contain antifreeze proteins, which are used in the food and aquaculture sectors, among others, in native or modified forms for commercial purposes. The manufacture of vaccines using cryo-preservatives generated from the extremophile Gelidibacter algens and tardigrades is the subject of extensive studies on cryptobiosis, or resistance to freezing, desiccation, or oxygen shortage (Ramakrishnan and Nalini, 2023).

7.2 Fish AFPs

Scholander and DeVries initially observed that certain fish inhabiting polar regions could withstand extremely cold water temperatures, which may occasionally dip below freezing.\ (Scholander, 1957; DeVries and Wohlschlag, 1969; DeVries, 1971). They attempted to determine how these fish managed to endure cold water, which was lower than the freezing point of fish blood, after making this discovery. The formation of ice crystals was shown to be slow and delayed when chilling was further increased, possibly because glycoproteins lower the non-colligative freezing point of solutions (Komatsu *et al.*, 1970). The names of the AFGPs are given to these proteins. Subsequently, Pseudopleuronectes americanus, a winter flounder, was found to contain no glycosylated AFPs (type I AFPs) (Duman and DeVries, 1974). Furthermore, several AFP types in the Arctic and Antarctic have been identified and categorized into four distinct categories (I, II, III, and IV). Although they exhibit fundamentally different primary sequences and three-dimensional structures, all types of AFP share similar characteristics that allow them to bind to ice and reduce the freezing point of solutions. Moreover, there is no evidence to suggest that any of these AFP subtypes share common ancestral genes.

The repeating sequence motif found in AFGPs, consisting of three amino acids (Ala-Ala-Thr), is connected to the hydroxyl group of the threonine residue by a disaccharide (Slaughter 1981). However, variations in the sequence occur at the first residue position, with Pro, Thr, or Arg occasionally substituted for the initial Ala residue. The names of the eight AFGPs (AFGP1–8) were determined by the number of units they contained. AFGP1 has the highest molecular weight (33.7 kDa) because it consists of approximately 50 repeating units, whereas AFGP8 has the lowest molecular weight (2.6 kDa), with only four repeating units. The antifreeze properties of AFGPs are often dependent on the number of repeating units. It is believed that AFGPs with higher molecular weights cover a larger surface area of ice and are more effective at limiting ice formation than smaller AFGPs (Feeney and Yeh, 1978; Knight *et al.*, 1984; Kao *et al.*, 1986).

Additionally, recent research has demonstrated the significance of carbohydrate moieties in AFGP activity. Nuclear magnetic resonance (NMR) structural investigations have shown that Ala residues and carbohydrate moieties are on different sides of the molecule. AFGPs have helical forms and amphipathic properties. As a result, the AFGPs exhibited significant recrystallization capabilities. There are a few restrictions regarding their commercial application for cryopreservation. Natural polar fish sources are insufficient to manufacture significant amounts of AFGPs, and it is challenging to build chemical synthesis systems for large-scale mass production. Conversely, AFPs can be produced in vast amounts using recombinant protein expression methods. Therefore, application studies have employed AFPs more frequently than they have employed AFGPs (Kim *et al.*, 2017).

7.3 Diatom AFPs

Novel AFP genes have been found in polar sea diatoms such as Berkeleya sp., Nitzschia frustulum, Fragilariopsis sp., and Chaetoceros neogracile. Furthermore, research on gene expression has revealed that stressors such as high salinity and low temperatures can alter AFP gene expression. Thus, AFP genes may be essential for the capacity of diatoms to adapt to their environments. The antifreeze action of recombinant antifreeze protein (Cn-AFP), which was first produced in 2009 from the oceanic diatom C. neogracile by Gwak *et al.*, was described (Gwak *et al.*, 2009).

7.4 Bioprospecting Pros

The discovery of novel chemical compounds in natural resources can be greatly aided by bioprospecting. It has aided businesses in creating, marketing, and releasing goods that benefit people, while also generating significant financial gains. Below is a list of the advantages of bioprospecting:

7.5 Finding significant genes: One of the primary benefits of bioprospecting is the search for significant genes. Bioprocessing-discovered genetic information can be used in several ways. Genetic material extracted from plants, including the opium poppy and Madagascar periwinkle, is an example.

7.6 Increased comprehension of the use of biological resources: Many think that biological resources will be better protected and cared for as more is learned about them and their uses. In regions where bioprocessing is more prevalent, preservation of natural resources is more likely to occur.

7.7 Treatment accessibility: There are more alternatives for treating various illnesses and other disorders when new natural resources are available. These choices offer additional means by which individuals might achieve health and overcome their illnesses.

8. Bioprospecting Cons

Although bioprospecting is generally seen as a beneficial technique that helps many people, it may have some drawbacks. There are a few drawbacks to bioprospecting.

8.1 Conservation: Preserving and protecting natural resources with caution is advocated by many scientists. The process of bioprospecting involves looking for resources in natural environments that are advantageous to people. The excessive use of nature poses a problem because it discourages conservation.

8.2 Economics: The search for commercially useful products is one of the main drawbacks of bioprospecting. Finding a useful human product that can be sold is the ultimate aim of bioprospecting. A resource that might help a limited number of people could be discovered, but there is a risk that no one would ever know about it, since commercializing it would not be particularly profitable.

8.3 Exploitation: Excessive resource utilization can result in environmental exploitation as well as the exploitation of local populations living in bioprospecting regions. For instance, the San tribe of Southern Africa has long used hoodia plants. They discovered that the plants effectively reduced hunger. The South African Council for Scientific and Industrial Research and Phyto Pharm, a business that collaborated with Pfizer, inked a contract in 1998 to commercialize plants for use as medications for weight reduction in the West. San people were not informed about the agreement or asked about their use or familiarity with the plant's advantages.

Conclusion

The exploration of biological diversity has unveiled a wealth of resources that hold the potential to address the pressing challenges facing the industry, such as overfishing, environmental degradation, and the demand for alternative protein sources. Understanding the diverse applications of bioprospecting from the development of novel aquafeeds to the identification of resilient and economically valuable fish species. The integration of biotechnology and genomics further accelerates the discovery and harnessing of valuable traits, propelling the industry towards enhanced productivity and environmental responsibility. Sustainable practices and biodiversity conservation must remain central to the bioprospecting framework to prevent potential negative consequences. Striking a balance between economic benefits and environmental responsibility is imperative for ensuring the long-term success of bioprospecting initiatives.

References

Achankunju, J. (2016). Bioprospecting selected microalgae for Aquaculture feed. ResearchGate. https://www.researchgate.net/publication/307637900_Bioprospecting_selected_microalgae_for_Aquaculture_feed

Akiwumi, P. How marine bioprospecting can spur development in small island states. (2022, June 23). UNCTAD. https://unctad.org/news/blog-how-marine-bioprospecting-can-spur-development-small-island-states

Arnaud-Haond, S., Arrieta, J. M., & Duarte, C. M. (2011b). Marine Biodiversity and gene patents. Science, 331(6024), 1521–1522. https://doi.org/10.1126/science.1200783

Arrieta, J. M., Arnaud-Haond, S., & Duarte, C. M. (2010). What lies underneath: Conserving the oceans' genetic resources. Proceedings of the National Academy of Sciences of the United States of America, 107(43), 18318–18324. https://doi.org/10.1073/pnas.0911897107

Bhooleshwari, Verma, DK., Maurya1, NK., & Mishra, SK. (2023). Bioprospecting and bio-indicator. Bioprospecting and Its Future, 1, 119.

Chakraborty, K. (n. d.). Marine bioprospecting for income and employment. Central Marine Fisheries Research Institute report.

Costello, M. J., Coll, M., Danovaro, R., Halpin, P., Ojaveer, H., & Miloslavich, P. (2010). A census of marine biodiversity knowledge, resources, and future challenges. PLOS ONE, 5(8), e12110. https://doi.org/10.1371/journal.pone.0012110

Cox PA, Balick MJ. (1994, June 1). The ethnobotanical approach to drug discovery. 270:82-87. PubMed. https://pubmed.ncbi.nlm.nih.gov/8023119/

Davies-Coleman, M. T., & Sunassee, S. N. (2012). Marine bioprospecting in Southern Africa. In Springer eBooks (pp. 193–209). https://doi.org/10.1007/978-3-642-28175-4_8

DeVries, A. L. (1971). Glycoproteins as biological antifreeze agents in Antarctic fishes. Science, 172(3988), 1152–1155. https://doi.org/10.1126/science.172.3988.1152

DeVries, A. L., & Wohlschlag, D. E. (1969). Freezing resistance in some Antarctic fishes. Science, 163(3871), 1073–1075. https://doi.org/10.1126/science.163.3871.1073

Duman, J. G., & DeVries, A. L. (1974). Freezing resistance in winter flounder Pseudopleuronectes americanus. Nature, 247(5438), 237–238. https://doi.org/10.1038/247237a0

Feeney, R. E., & Yeh, Y. (1978). Antifreeze Proteins from Fish Bloods. In Advances in Protein Chemistry (pp. 191–282). https://doi.org/10.1016/s0065-3233(08)60576-8

Global Ocean Commission (2013). Policy Options Paper No. 4: Bioprospecting and marine genetic resources in the high seas. A series of papers on policy options, prepared for the third meeting of the Global Ocean Commission. November 2013. Available at http://www.globaloceancommission.org/wp-content/uploads/GOC-paper04- bioprospecting.pdf

Gwak, I. G., Jung, W. S., Kim, S. H., Kang, S., & Jin, E. (2009). Antifreeze Protein in Antarctic Marine Diatom, Chaetoceros neogracile. Marine Biotechnology, 12(6), 630–639. https://doi.org/10.1007/s10126-009-9250-x

Harinkhede, H., Inwati, P., Verma, DK., & Khan, A. (2023). Bioprospecting for biodiversity. Bioprospecting and Its Future, 1, 58.

Kao, M. H., Fletcher, G. P., Wang, N. C., & Hew, C. L. (1986). The relationship between molecular weight and antifreeze polypeptide activity in marine fish. Canadian Journal of Zoology, 64(3), 578–582. https://doi.org/10.1139/z86-085

Katherin Sylvia. R. (2022). Bioprospecting and Bioentrepreneurship Potentials in Marine Biology - A Brief Overview. Dr. N. Yogananth, Dr. Sheeba E, Dr. T. Sivakumar, Dr. R. Bhakyaraj. (Eds), Bioentrepreneurship in Biosciences - Recent Approaches. India: Darshan Publishers. pp: 28-45. http://dx.doi.org/10.22192/bbra.2022

Kim, S. H., Lee, J. H., Hur, Y. B., Lee, C. W., Park, S., & Koo, B. (2017). Marine Antifreeze proteins: structure, function, and application to cryopreservation as a potential cryoprotectant. Marine Drugs, 15(2), 27. https://doi.org/10.3390/md15020027

Knight, C., DeVries, A. L., & Oolman, L. D. (1984). Fish antifreeze protein and the freezing and recrystallization of ice. Nature, 308(5956), 295–296. https://doi.org/10.1038/308295a0

Komatsu, S., DeVries, A. L., & Feeney, R. E. (1970). Studies of the structure of freezing point-depressing glycoproteins from an Antarctic fish. The Journal of biological chemistry, 245(11), 2909–2913.

Laird, S. (2013). Bioscience at a Crossroads: Access and Benefit Sharing in a Time of Scientific, Technological and Industry Change: The Pharmaceutical Industry. Secretariat of the CBD, Toronto

Leal, M. C., Munro, M. H. G., Blunt, J. W., Puga, J., Jesus, B., Calado, R., Rosa, R., & Madeira, C. (2013). Biogeography and biodiscovery hotspots of macroalgal marine natural products. Natural Product Reports, 30(11), 1380. https://doi.org/10.1039/c3np70057g

Leal, M. C., Puga, J., Serôdio, J., Gomes, N. C. M., & Calado, R. (2012). Trends in the Discovery of New Marine Natural Products from Invertebrates over the Last Two Decades – Where and What Are We Bioprospecting? PLOS ONE, 7(1), e30580. https://doi.org/10.1371/journal.pone.0030580

Marris, E. (2006). Drugs from the deep. Nature, 443(7114), 904–905. https://doi.org/10.1038/443904a

Martins, A. M., Vieira, H., Gaspar, H., & Santos, S. (2014). Marketed marine natural products in the pharmaceutical and cosmeceutical industries: Tips for success. Marine Drugs, 12(2), 1066–1101. https://doi.org/10.3390/md12021066

Molinski, T. F., Dalisay, D. S., Lievens, S. L., & Saludes, J. P. (2008b). Drug development from marine natural products. Nature Reviews Drug Discovery, 8(1), 69–85. https://doi.org/10.1038/nrd2487

Ostrander, G. K., Cheng, K. C., Wolf, J. C., & Wolfe, M. J. (2004b). Shark cartilage, cancer and the growing threat of pseudoscience. Cancer Research, 64(23), 8485–8491. https://doi.org/10.1158/0008-5472.can-04-2260

Pirzadah, T. B., Malik, B., Maqbool, T., & Rehman, R. U. (2019). Development of Nano-Bioformulations of nutrients for sustainable agriculture. In Nanotechnology in the life sciences (pp. 381–394). https://doi.org/10.1007/978-3-030-17061-5_16

Rajakumar, K. (2003). Vitamin D, Cod-Liver Oil, Sunlight, and Rickets: A Historical perspective. Pediatrics, 112(2), e132–e135. https://doi.org/10.1542/peds.112.2.e132

Ramakrishnan, D., & Nalini, D. (2023). Bioprospecting for ecological restoration. Bioprospecting and Its Future, 1, 88.

Rishton, G. M. (2008). Natural products as a robust source of new drugs and drug leads: past successes and present day issues. The American Journal of Cardiology, 101(10), S43–S49. https://doi.org/10.1016/j.amjcard.2008.02.007

Rollo, A. J. & Cheprasov, A. (2023). Bioprospecting Definition, Pros & Cons. study.com. Retrieved November 21, 2023, from Bioprospecting Definition, Pros & Cons - Video & Lesson Transcript | Study.com

Rosendal, G. K. (2006). Regulating the use of genetic resources – between international authorities. European Environment, 16(5), 265–277. https://doi.org/10.1002/eet.424

Ryan, J. T., Ross, R. P., Bolton, D., Fitzgerald, G. F., & Stanton, C. (2011). Bioactive Peptides from Muscle Sources: Meat and Fish. Nutrients, 3(9), 765–791. https://doi.org/10.3390/nu3090765

Scholander, P. F., Dam, L., Kanwisher, J., Hammel, H. T., & Gordon, M. (1957). Supercooling and osmoregulation in arctic fish. Journal of Cellular and Comparative Physiology, 49(1), 5–24. https://doi.org/10.1002/jcp.1030490103

Silva, T. H., Alves, A., Ferreira, B., Oliveira, J. M., Reys, L. L., Ferreira, R., Sousa, R. A., Silva, S. S., Mano, J. F., & Reis, R. L. (2012). Materials of marine origin: a review on polymers and ceramics of biomedical interest. International Materials Reviews, 57(5), 276–306. https://doi.org/10.1179/1743280412y.0000000002

Slaughter, D., Fletcher, G. L., Ananthanarayanan, V. S., & Hew, C. L. (1981). Antifreeze proteins from the sea raven, Hemitripterus americanus. Further evidence for diversity among fish polypeptide antifreezes. The Journal of biological chemistry, 256(4), 2022–2026.

Verma, DK., Inwati, P., & Harinkhede, H. (2023). Bioprospecting in marine aquaculture. Bioprospecting and Its Future, 1, 188.

7

Minimal Water Usage System in Aquaculture

Kinjal J. Patel **and** ***Ketan V. Tank***

Department of Aquaculture, College of Fisheries Science, Kamdhenu University Veraval-362266, Gujarat, India

Abstract

The sustainable management of water resources is paramount in modern aquaculture practices to mitigate environmental impacts and ensure long-term viability. This chapter explores innovative methods and innovations meant to reduce water consumption in aquaculture systems. We start with a summary of the significance of water conservation in aquaculture before going into several strategies including aquaponics, biofloc technology, and recirculating aquaculture systems (RAS). Recirculating aquaculture systems (RAS) consume less water and area while producing fish in large quantities. A balance between water use, waste discharge, energy consumption, and productivity must be found to accomplish both environmentally friendly and economically viable production. The biofloc systems emphasise the significance of heterotrophic bacteria in transforming organic materials into microbial aggregates that function as a source of nutrients and a mechanism for water treatment. The creation of microbial biomass is encouraged by biofloc systems, which also maintain excellent water quality by means of carbon and nitrogen cycling. These benefits include less water exchange needs, increased resistance to disease, and increased feed conversion efficiency. In a symbiotic system, hydroponically grown plants are fertilized by fish excrement, while plants filter water for fish, resulting in aquaponics. This closed-loop method uses less water, doesn't require artificial fertilizers, and reduces environmental impact.

Keywords*: Recirculating aquaculture systems (RAS), biofloc, aquaponics.*

1. Introduction

High animal stocking densities and minimal water exchange are characteristics of an intensive cultivation system. Aquaculture farms can utilize as much as 45 cubic meters of water for every kilogram generated in ponds. Water consumption in the current aquaculture pond system can be decreased by (1) choosing feed ingredients with low water requirements, (2) improving feed production within the system, and (3) combining aquaculture and agriculture.

However, pond aquaculture will not become more water-efficient with these techniques. Only indoor fish farming techniques, such as aquaponics, biofloc technology (BFT), recirculating aquaculture, and other zero-water exchange aquaculture production systems, can do this (Jusoh *et al.*, 2020). it is simply the combined culture of fish and plants in a recirculating system. Mainly, recirculatory aquaculture systems are designed to raise large quantities of fish in relatively small volumes of water by treating the water to remove toxic waste products and reuse again for fish culture (Rakocy & Hargreaves, 1993; Timmons *et al.*, 2010; Verdegem, 2013).

2. Recirculatory Aquaculture System (RAS)

Recirculating aquaculture systems (RAS) consume less water and area while producing fish in large quantities. With the Recirculatory Aquaculture System (RAS), suspended particles and metabolites are removed from water through mechanical and biological filtration, and then the water is recycled and reused. This technique uses the least amount of water and land possible to cultivate fish at high densities across a variety of species. It is not like other aquaculture production techniques in that it is an intense, high-density fish culture.

With this technology, fish are usually raised in indoor/outdoor tanks in a controlled environment as opposed to the conventional practice of growing fish outside in open ponds and raceways (Tom *et al.*, 2021). Water is filtered and cleaned by recirculating systems and then recycled back into fish culture tanks. Any species that is raised in aquaculture can be processed using this technology, which is based on the use of mechanical and biological filters. Only enough fresh water is added to the tanks to compensate for evaporation, splash out, and waste material flushing. The system is filled with reconditioned water, and every day, no more than 10% of the system's total water volume is added. The fish farmer must grow as much fish as possible within the built-in capacity to compete competitively and make the most use of the significant financial investment in the recirculating system (Funge-Smith *et al.*, 2001).

Recirculation aquaculture systems (RAS) recycle the treated water after eliminating harmful contaminants from wastewater. Thus, comparatively little water is used in RAS systems to produce a large number of fish. Recirculating wastewater to the fish tanks might be done in part or in whole. Depending on the supply and feeding rates, recirculating aquaculture systems typically replace 5 to 10% of their water each day (Zhang *et al.*, 2011). Recirculating aquaculture systems (RASs) are environmentally benign since they minimise the impact of fish production on pollution and freshwater consumption (Helfrich *et al.*, 1991). Debris removal, organic matter removal, ammonia removal, and nitrite removal are essential for developing recirculating aquaculture systems.

The type of filtration and the amount and quality of feed are critical components in the management of recirculating systems. Recirculating systems employ a variety of filter types, but the ultimate objective of all filtration is to provide aquatic creatures with high-quality water by eliminating solids, surplus nutrients, and metabolic wastes. It is crucial to take all aspects into account while developing and funding aquaculture systems. However, it is recommended to encourage Backyard Recirculation Aquaculture Systems to support small-scale fish farmers and entrepreneurs as well as to facilitate fish production in urban and semi-urban locations where land and water are minimal (Farghally *et al.*, 2014).

2.1 How does RAS operate?

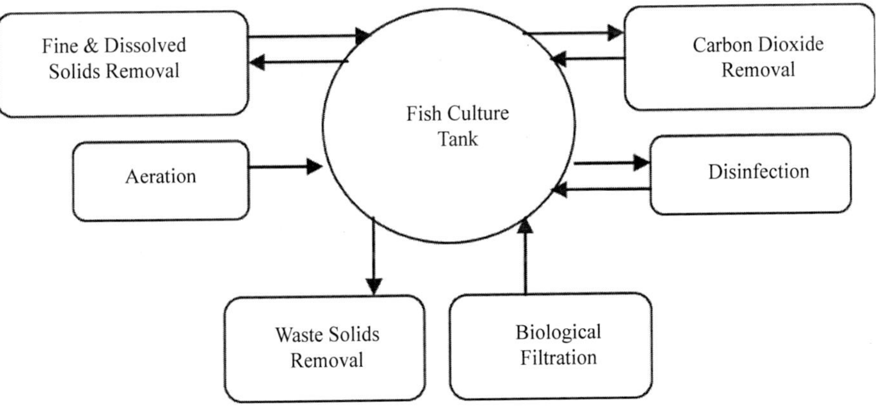

Fig. 1. Recirculation aquaculture systems (RAS) components. (Farghally *et al.*, 2014)

2.2 Benefits of RAS

- Provide the ability to produce a wide variety of species regardless of the required temperature.
- When feeding can be closely watched for a full day, feed management in RAS is much improved.
- It is possible to mitigate the stress that stocks experience from various reasons, including unfavorable weather, unfavorable temperature conditions, external pollutants, and predation.
- Facilitate the safe cultivation of non-endemic species.

- Extended equipment and tank durability
- A lower reliance on antibiotics and other medications, which results in the benefit of obtaining premium fish. reduction in the direct operating expenses related to parasites, feed, and predator control.
- Decreased risk as a result of disease, parasites, and climate RAS production can encourage farming location flexibility and market closeness.
- Possibly stop parasites from escaping into recipient rivers and lakes.

2.3 A negative aspect of RAS

- It requires a continuous, uninterrupted power source; if the electricity fails, backup power is needed.
- The initial investment in a recirculating aquaculture system is more than that of ponds and raceways.

2.4 Species suitable for RAS

- Silver/Indian Pompano (Trichinotus Blochii/ Trichinotus mookalee)
- Pangasius (Pangasianodon hypophthalmus)
- Cobia (Rachycentron canadum)
- Baramundi/ Asian Seabass/Bhetki (Lates calcarifer)
- Tilapia (Oreochromis niloticus)
- Pearl spot/Karimeen (Etroplus suratensis)
- Rainbow Trout (Oncorhynchus mykiss)

2.5 RAS components include

- Insulated shed/ Building
- Store cum office for feed and accessories
- Pump house Grow out tanks: Circular cement tanks/FRP tanks, including inlet, outlet central drainage ã Settling tanks for sludge
- Water Storage (sump) tanks
- Overhead tanks.
- Mechanical (Hydraulic) filters, Drum filters, Glass wool/ muslin cloth filter
- Pumps and motors
- Power generator
- Sludge collector, solid collectors
- Biofilters, UV units

- Electrification
- Automatic feeder (wherever required)
- Aeration system (air/ oxygen), Carbon dioxide trapper system (degasser)
- Water testing kit
- Water supply system, bore well, etc. (wherever required)
- Inputs such as Seed, Feed, additives and supplements, electricity/ Diesel, manpower, etc.

2.6 Feed

- A diet rich in protein that includes all the necessary vitamins and minerals Depending on the quality and protein level of the meal, feeding should be done at a rate of 3–5% of the fish's body weight. Regular feedings, multiple times a day, will lead to higher growth rates and an enhanced feed conversion ratio.

3. Biofloc Fish Culture

Biofloc Technology (BFT) is considered a new “blue revolution” since nutrients can be continuously recycled and reused in the culture medium, benefiting from the minimum or zero-water exchange. BFT is an ecologically conscious aquaculture method that uses the generation of microorganisms in situ. The suspended growth known as "bio floc" in ponds and tanks is made up of aggregates of bacteria, phytoplankton, and other grazing bacteria as well as dead and living particle organic debris (Nasir *et al.*, 2015). The procedure involves using the microbial activities present in the pond or tank to clean the water and supply food resources for the cultivated organisms (Jatoba *et al.*, 2019). As a result, other names for this system include heterotrophic ponds, green soup ponds, and active suspension ponds.

3.1 How does BFT operate?

The use of biofloc systems for wastewater treatment in aquaculture has become increasingly important.

The technique's basic idea is to maintain a greater C-N ratio by supplying carbohydrates and it also improves the quality of the water by producing high-grade single-cell microbial protein.

Under such circumstances, heterotrophic microbial growth takes place, assimilating nitrogenous waste that can be used as feed by the cultured species and functioning as a bioreactor to control the quality of the water (Yu *et al.*, 2023). Because heterotrophs grow at a rate ten times faster than autotrophic nitrifying bacteria and produce more microorganisms per unit of substrate, harmful nitrogen species are fixed more quickly in biofloc.

The basic concept of this technique is the idea of flocculation inside the system (Oliveira *et al.*, 2022).

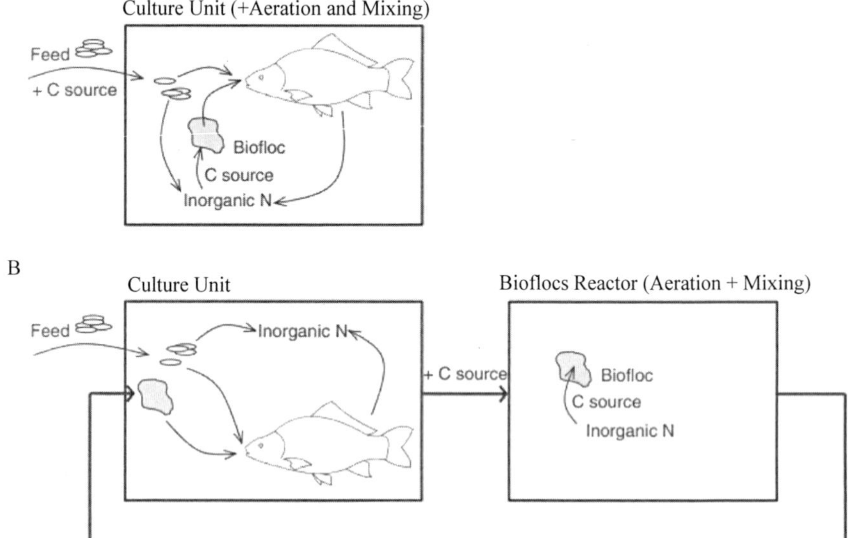

Fig. 2. Biofloc technology(A) Integration of bioflocs within the culture unit by using feed with a relatively low N content. (B) Use of a separate bioflocs reactor.

3.2 Species suitable for Biofloc Culture

Some of the species that are suitable for BFT are:

Air-breathing fish like Singhi (Heteropneustes fossilis), Magur (Clarias batrachus), Pabda (Ompok pabda), Anabas/Koi (Anabas testudineus), Pangasius (Pangasianodan hypophthalmus)

Non-air-breathing fishes like Common Carp (Cyprinus carpio), Rohu (Labeo rohita), Tilapia (Oreochromis niloticus), Milkfish (Chanos chanos) Shellfishes like Vannamei (Litopenaeus vannamei) and Tiger Shrimp (Penaeus monodon)

3.3 The nutritional value and composition of Biofloc

A heterogeneous mixture of suspended particles and different microorganisms connected to extracellular polymeric materials is called biofloc. Micro-organisms including bacteria, fungi, algae, invertebrates, and detritus, among others, make up this substance.

It is a live feed that is high in protein that is created when leftover feed and excrement in a culture system are exposed to sunlight and intense aeration, turning them into a natural food.

The loose mucus matrix that binds each floc together is released by bacteria and is held together by electrostatic attraction or filamentous microorganisms (Oliveira *et al.*, 2022). The naked eye can see large flocs, but the majority are tiny. The size range of floc is 50–200 microns.

Biofloc has a good nutritional value. Protein has a dry weight range of 25–50% and fat has a dry weight range of 0.5–15%. It is an excellent source of minerals and vitamins, especially phosphorus. It functions similarly to probiotics. It is suggested that dried biofloc be used in place of fishmeal or soybeans in the feed (Crab *et al.*, 2012).

3.4 Advantage of BFT

- It reduces environmental impact
- Judicial use of land and water
- Higher productivity (It enhances survival rate, growth performance, and better feed conversion in the culture systems of fish)
- Higher biosecurity.
- Reduces water pollution and mitigates the risk of the introduction and spread of pathogens
- Eco-friendly culture system.
- Limited or zero water exchange system
- It reduces the utilization of protein-rich feed and the cost of standard feed
- It reduces the pressure on capture fisheries i.e., use of cheaper food fish and trash fish for fish feed formulation.

3.5 How Is the Inoculum Prepared?

Method: 150 liters of inoculum are needed for the floc formation in 15000 liters of fresh water.

Step 1: Fill a clean tub or can with 150 liters of water, then quickly aerate it.

Step 2: Add 30 grams of carbon source (tapioca flour, wheat flour, or Jarry) along with 3 kg of pond soil and 1.5 grams of ammonium sulfate or urea.

Step 3: Thoroughly combine with the water in the tub and ensure sufficient aeration.

Step 4: The inoculum can be moved to main tank once it is ready, which should happen in 24 to 48 hours.

Daily addition of carbon source is required for the development of floc.

4. Aquaponics System

The combined technique of fish and plant production known as aquaponics is mainly made up of two subsystems: hydroponics and aquaculture. The basic idea is to divide and share nutrient resources between fish and plants, as well as to effectively use water to grow two crops instead of one. This integrated farming approach is widely utilized to cultivate fish and vegetables in urban settings with limited resources. Aquaponics is the culture of fish and horticultural plants (Ravindranath *et al.*, 2017). Many plants can be used in aquaponics systems; however, which ones are best for a given system will depend on the fish's stocking density and age. Aquaponics systems are a good fit for green leafy vegetables with low to medium nutrient requirements, such as tomatoes, capsicum, lettuce, cabbage, lettuce, basil, spinach, chives, and herbs.

Aquaponics equipment mainly includes vegetable cultivation and fish culture equipment (Reshmi Menon *et al.*, 2013). The vegetable cultivation equipment mainly introduces three vegetable planting methods in common facilities with hydroponics as the main method: the nutrient film technique (NFT), the deep flow technique (DFT), and the floating capillary hydroponics (FCH).

List of plants that grow best on a specific aquaponics method:

Media Based Aquaponics System: Lettuce, tomatoes, ginger, eggplant, cucumber, and any plants that will fit your grow bed.

Raft System: Lettuce, basil, kale, cabbage, Swiss chard, bok choy, mint, watercress, and other small rooting plants

Nutrient Film Technique (NFT): Lettuce, strawberry, spinach, parsley, dill, and other small rooting plants.

It is a Recirculation Culture System in which fish waste is discharged into bio-filter troughs containing horticultural plants and fish are fed high-quality floating pellet feed (Wei *et al.*, 2019). The water flow rate is controlled with the use of a timer. Aquaponics systems grow entirely organic fish and plants. The system has a significant initial investment, but it has low recurrent costs and yields respectable profits. One benefit of this system is that it uses less water, less land, recycles trash, requires less manpower, etc. In an aquaponic system, animals and plants coexist in a symbiotic connection.

4.1 How Does Aquaponics Work?

In aquaponics, the plants are grown in the grow bed, and fish are placed in the fish tank. The nutrient-rich water from the fish tank that contains fish waste is fed to the grow bed, where billions of naturally occurring beneficial bacteria break the ammonia down into nitrites and then into nitrates. Plants absorb these nitrates and other nutrients to help them grow. The plant's roots clean and filter the water before it flows back into the fish tank for the fish to live. The fresh, clean, and oxygenated water recirculates back to the fish tank, where the cycle will begin again (Perla M.F. *et al.*, 2015).

In this farming system, fish eat the food and release trash which is transformed by useful bacteria into nutrients. plants can use these nutrients. And these nutrients plants help to clarify the water. Thereafter, the ammonia-rich water flows from the fish tank into a biofilter with un-eaten food and decaying plant matter together. After that, bacteria break everything down into organic nutrient solutions inside the biofilter, just for growing vegetables.

Aquaponic freshwater systems include three main components- freshwater fish, nitrifying bacteria, and plants, they depend on each other to survive. Without bacteria, plants wouldn't have the functional form of nutrients to eradicate the fish waste either, which is why natural filtration is crucial.

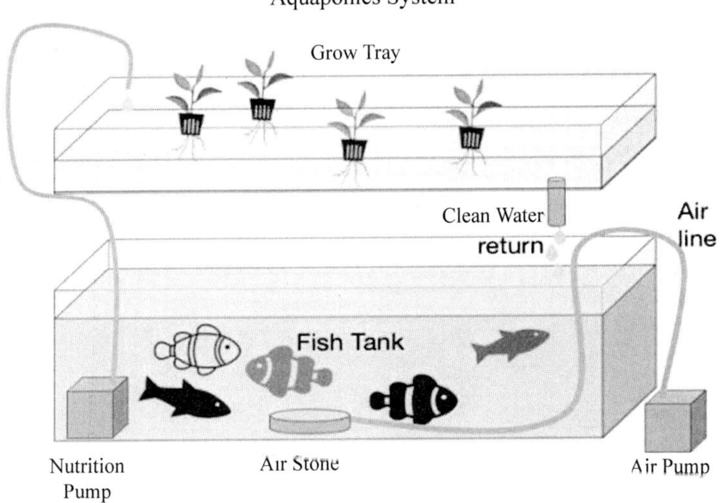

Fig. 3. Aquaponic system

4.1.1 Resources

One common prediction for the future of food production is aquaponics. Without the use of dangerous chemicals or pesticides, aquaponic systems are claimed to use only 2 to 10% of the water needed for conventional vegetable or crop production, with the potential to produce 10 times the yield. The most important feature of aquaponics is how little land or space it requires, which gives rise to concepts like "urban agriculture", "urban farming", "urban gardening," "terracing," "vertical gardening," "office farming" (indoor), etc.

Because natural resources are highly efficient, they are used very sparingly, which preserves priceless natural resources like water, land & the environment.

4.1.2 Status and Potential

In our nation, aquaponics is a relatively new technique. Therefore, the creation of these units will enhance fish farmers' understanding of current and upcoming aquaculture technologies. The biology of fish cultivated in these units, technology, and strict water quality criteria make the operation of these units more difficult.

Our nation leads the world in the production of freshwater fish because farmers have harvested anywhere from 2 to 10 tonnes of fish per acre annually using even very basic techniques of fish farming. In addition to producing a healthy crop of vegetables, an aquaponics system may grow up to five times as many fish in the same space each year.

4.2 Project Location and Execution

4.2.1 Site Selection: Although aquaponic systems work well in situations where there isn't enough water to remove fish waste from the production system, choosing a proper location is crucial. Ammonia and other waste products are eliminated from water by passing it through a treatment unit; this has the same effect as a flow-through design. Construction requires a minimum of 150 m^2 of land or space for a backyard-style aquaponics system and 2000 m^2 for a small-scale commercial aquaponics system (Christopher Somerville *et al.*, 2014).

4.2.2 Applicants: Applicants comprise female self-help groups, fisherman's organizations, fish farmers, and entrepreneurs; selection would be predicated on their level of awareness and interest.

4.2.3 Project Implementation: The beneficiary will carry out the project with technical assistance from the State Government's Department of Fisheries and the Designated Technology/Service Provider. The government (federal/state) will provide financial help in the form of a subsidy; the beneficiary will be responsible for covering the remaining cost through bank loans, self-financing, etc. (M.N. Mamatha and S.N. Namratha, 2017).

4.3 Project Components

4.3.1 Water Quality: Water quality is important and the optimum range of certain parameters is required for successful fish culture in an Aquaponics System are as follows:

S.No.	Water Parameter	Optimum Range
1.	Temperature	26-30 °C
2.	Dissolved Oxygen	4-6 ppm
3.	pH	7-8
4.	Alkalinity	120-150 ppm
5.	Ammonia	< 0.05 ppm
6.	Nitrite	< 0.5 ppm
7.	Nitrate	< 5 ppm

4.3.2 Targeted Fish Species: Tilapia (Oreochromis niloticus), Pangasius (Pangasiandon hypophthalmus) or any species that can tolerate high-density stocking are suitable for the Aquaponic System.

Ex.

GIFT Tilapia (Oreochromis niloticus)

Pangasius (Pangasiandon hypophthalmus)

There are several advantages and disadvantages of this aquaponics farming system.

4.4 Advantages of Aquaponics

The financial and environmental advantages of this system such as healthy and organic benefits, and reduced water uses are essential elements for tolerable farming.

4.4.1 Health Benefits & Organic: Growing food in an aquaponic garden is more nutrient-dense, revitalising, and authentically organic. Plants and fish are not contaminated by biopesticides. Everything that grows fruits and vegetables in this farming system is chemical-free and natural (S. Selvi and U. Vanitha, 2012).

4.4.2 Decreases Water Usage: In comparison to commercial farms and backyard gardens, water waste is greatly reduced. Water is often used, recovered, and reused to operate an aquaponic system. Additional water is not required to sustain the growth of plants or fish.

4.4.3 Environmental Benefits: Of this growing system, the environmental benefits are the most significant. This agricultural method uses very little power in general and extremely little water. This system is closed. Both trash and pollution are absent.

4.4.4 Chemical-Free: No chemicals or pesticides of any kind that could harm the fish are needed. There are several advantages for the environment and human health when chemicals are not used.

4.4.5 Climate Adaptive: As the world's population grows and food insecurity rises, climate adaption becomes more crucial. It's a true climate-adaptive growing technique that lets farmers make changes and conserve energy.

4.4.6 Commercial Aquaponics Farmers Earn Two Incomes: A bigger aquaponic farming system can provide two different kinds of revenue. They can also sell the fish, fruits, and vegetables they cultivate to food makers (C Somerville *et al.*, 2014).

4.5 Disadvantages of Aquaponics

The aquaponics farming technique has numerous benefits, but it also has certain disadvantages. Setting up the growth system can be costly. Certain mid-size producers may be prohibited by certain restrictions.

4.5.1 It is expensive: Setting up and maintaining an aquaponic growth system is expensive. The size of the aquaponic tank usually determines the majority of the costs.

4.5.2 This method is not available for certain crops: Since root crops need soil to develop, they are not growing in this system. Additionally difficult to grow with this strategy are large crops, assuming higher water and nutrient content. Additionally, several fish are not recommended for this developing system, including largemouth bass, yellow perch, salmon, and trout.

4.5.3 It must be lodged professionally: Professional skill and experience are required for the setup and lodging of the expanding system. Since keeping the plants carries some danger of financial loss, the care should be provided by professionals.

4.5.4 It wastes a lot of electricity: The growing technique could reduce the need for water and land. However, it balanced out its usage of electricity. Even when not a lot of energy is required, the energy is still necessary.

4.5.5 There's a chance of random failure: Aquaponics is thought to be a complex procedure since you have to closely monitor both the fish and the plants. If the conditions in which fish live change, they might not be able to stay.

5. Commercial Aquaponics System

A small-scale aquaponics unit that would be operated on a commercial basis would be best established on 0.5 acres (2000 square meters) of land. In addition to the Moving Bed Biofilm Reactor (MBBR), it would consist of one rectangular fish tank, ten plant grow beds, and various other equipment like filtration units, pumps, and aerators. Roughly Rs. 3.7 lakh would be the capital cost and Rs. 4.1 lakh would be the operating expenditures, for a total estimated cost of Rs. 7.8 lakh (Okomoda *et al.*, 2023).

Conclusion

The implementation of minimal water usage systems in aquaculture, such as recirculating aquaculture systems (RAS) and biofloc technology, represents a significant step towards achieving sustainability in the industry. These systems minimize environmental impact, minimize resource use, and improve operational efficiency by optimizing water utilization. Aquaculture practitioners may enhance resilience and long-term sustainability in a fast-changing global setting, while simultaneously mitigating water-related difficulties through inventive strategies and technical improvements. Adopting these tactics encourages a paradigm change towards conscientious water management, guaranteeing aquaculture's sustained prosperity while preserving aquatic habitats for future generations. certainly, an aquaponics production system that combines a recirculating aquaculture system with soilless plant cultivation is an environmentally sustainable choice for food production in the twenty-first century. This would entail integrating the aquaponics system with other food production systems, which could lead to improved productivity and efficiency, less waste to be disposed of, and lower energy and water use.

References

C Somerville, M Cohen, E Pantanella, A Stankus and A Lovatelli, Small-Scale Aquaponic Food Production, Integrated Fish and Plant Farming, FAO, Fisheries and Aquaculture Technical Paper, 2014.

Christopher Somerville, Moti Cohen, Edoardo Pantanella, Austin Stankus and Alessandro Lovatelli, 2014. "Small-scale aquaponic food production – Integrated fish and plant farming". FAO Fisheries and Aquaculture Technical Paper 589, 2014, pp 288.

Crab, R., Defoirdt, T., Bossier, P., & Verstraete, W. (2012). Biofloc technology in aquaculture: beneficial effects and future challenges. Aquaculture, 356, 351-356.

Farghally, H. M., Atia, D. M., El-Madany, H. T., & Fahmy, F. H. (2014). Control methodologies based on geothermal recirculating aquaculture system. Energy, 78, 826-833.

Funge-Smith, S., and Phillips, M. J. (2001). Aquaculture systems and species. In R. P. Subasinghe, P. Bueno, M. J. Phillips, C. Hough, S. E. McGladdery, & J. R. Arthur (Eds.), Aquaculture on the Third Millennium. Technical Proceedings of the Conference on Aquaculture in the Third Millenium, Bangkok, Thailand (pp. 129-135). NACA, Bangkok and FAO, Rome: Food and Agriculture Organization of the United Nations.

Helfrich, L. A., & Libey, G. S. (1991). Fish farming in recirculating aquaculture systems (RAS). Virginia Cooperative Extension.

Jatobá, A., Borges, Y. V., & Silva, F. A. (2019). BIOFLOC: sustainable alternative for water use in fish culture. Arquivo Brasileiro de Medicina Veterinária e Zootecnia, 71, 1076-1080.

Jusoh, A., Nasir, N. M., Yunos, F. H. M., Jusoh, H. H. W., & Lam, S. S. (2020, January). Green technology in treating aquaculture wastewater. In AIP Conference Proceedings (Vol. 2197, No. 1). AIP Publishing.

M.N.Mamatha and S.N.Namratha, "Design & implementation of indoor farming using automated aquaponics system," 2017 IEEE International Conference on Smart Technologies and Management for Computing, Communication, Controls, Energy and Materials (ICSTM), Chennai, India, 2017, pp. 396-401, doi: 10.1109/ICSTM.2017.8089192.

Nasir, N. M., Bakar, N. S. A., Lananan, F., Abdul Hamid, S. H., Lam, S. S., and Jusoh, A. Bioresource Technology, 190, 492-498 (2015).

Okomoda, V. T., Oladimeji, S. A., Solomon, S. G., Olufeagba, S. O., Ogah, S. I., & Ikhwanuddin, M. (2023). Aquaponics production system: A review of historical perspective, opportunities, and challenges of its adoption. Food science & nutrition, 11(3), 1157-1165.

Oliveira, L. K., Wasielesky Jr, W., & Tesser, M. B. (2022). Fish culture in biofloc technology (BFT): Insights on stocking density carbon sources, C/N ratio, fish nutrition and health. Aquaculture and Fisheries.

Perla M.F., Oscar A.J., Enrique R.G., et al., "Perspective for aquaponic systems: "Omic" technologies for microbial community analysis," BioMed Research International, vol. 2015, article id 480386, 10 pages, 2015.

Rakocy, J. E., & Hargreaves, J. A. (1993). Integration of vegetable hydroponics with fish culture: A review. In J.K. Wang (Ed.), Techniques for modern aquaculture, proceedings aquacultural engineering conference (pp. 112–136). American Society of Agricultural Engineers.

Ravindranath, K., 2017. Aquaponics – an Integrated Fish and Plant Production System for Urban, Suburban and Rural Settings. NFDB Newsletter Matsya Bharat, Vol. 8, Issue 5, January-March 2017.

Reshmi Menon and Shahana G.V and Sruthi V, Small Scale Aquaponics System, International Journal of Agriculture and Food Science Technology, 2013, 4 (10), 970-980.

S Selvi and U Vanitha, Organic Farming: Technology for Environment-Friendly Agriculture, International Conference on Advances in Engineering, Science and Management, 2012, 132 -136.

Timmons, M. B., Ebeling, J. M., Wheaton, F. W., Summerfelt, S. T., & Vinci, B. J. (2010). Recirculating aquaculture (2nd ed., pp. 906). Ithaca NY, USA. NRAC Publication.

Tom, A. P., Jayakumar, J. S., Biju, M., Somarajan, J., & Ibrahim, M. A. (2021). Aquaculture wastewater treatment technologies and their sustainability: A review. Energy Nexus, 4, 100022.

Verdegem, M. C. J. (2013). Nutrient discharge from aquaculture operations in function of system design and production environment. Reviews in Aquaculture, 5, 158– 171.

Wei, Y., Li, W., An, D., Li, D., Jiao, Y., & Wei, Q. (2019). Equipment and intelligent control system in aquaponics: A review. Ieee Access, 7, 169306-169326.

Yu, Y. B., Choi, J. H., Lee, J. H., Jo, A. H., Lee, K. M., & Kim, J. H. (2023). Biofloc technology in fish aquaculture: A review. Antioxidants, 12(2), 398.

Zhang, S. Y., Li, G., Wu, H. B., Liu, X. G., Yao, Y. H., Tao, L., & Liu, H. (2011). An integrated recirculating aquaculture system (RAS) for land-based fish farming: The effects on water quality and fish production. Aquacultural Engineering, 45(3), 93-102.

8

Precision Fisheries The Future of Farming

Kamin Alexander **[*]** ***and Shruti Samson***

Department of Biological Sciences, Sam Higginbottom University of Agriculture Technology and Sciences, Prayagraj, Uttar Pradesh (211007), India

Abstract

Precision fisheries, a concept adapted from precision agriculture, involves the application of advanced technologies and data-driven approaches to optimize fishing operations and resource management. This chapter explores the potential of precision fisheries as the future of fishing, offering benefits such as increased efficiency, reduced environmental impact, and improved sustainability. By integrating cutting-edge technologies such as remote sensing, GIS (Geographic Information Systems), IoT (Internet of Things), and machine learning, precision fisheries aim to enhance decision-making, maximize yield, and minimize resource wastage. The chapter discusses key components of precision fisheries, including real-time monitoring, predictive modeling, and targeted harvesting, and examines their implications for the fishing industry, ecosystems, and coastal communities. Furthermore, the chapter highlights challenges and opportunities associated with the adoption of precision fisheries and provides recommendations for policymakers, researchers, and stakeholders to harness its full potential for sustainable fisheries management and food security.

1. Introduction

Fishing, as one of the oldest human activities, has evolved significantly over millennia. From primitive methods like hand gathering to modern industrial fishing fleets, the fishing industry has undergone remarkable transformations driven by technological advancements and increasing global demand for seafood. However, conventional fishing practices have often led to overexploitation of marine resources, habitat destruction, and ecosystem degradation, raising concerns about the sustainability of fisheries worldwide. In response to these challenges, precision fisheries have emerged as a promising approach to revolutionize fishing practices and ensure the long-term viability of marine ecosystems and coastal communities.

Precision fisheries, inspired by the principles of precision agriculture, leverage cutting-edge technologies and data-driven methodologies to optimize fishing operations, improve resource management, and minimize environmental impact. By harnessing the power of remote sensing, GIS, IoT, and machine learning, precision fisheries enable real-time monitoring of fish stocks, ocean conditions, and vessel movements, allowing fishermen to make informed decisions and adapt their strategies accordingly. Moreover, predictive modeling techniques facilitate the identification of fishing hotspots, optimal harvesting times, and species distribution patterns, enhancing the efficiency and sustainability of fishing activities

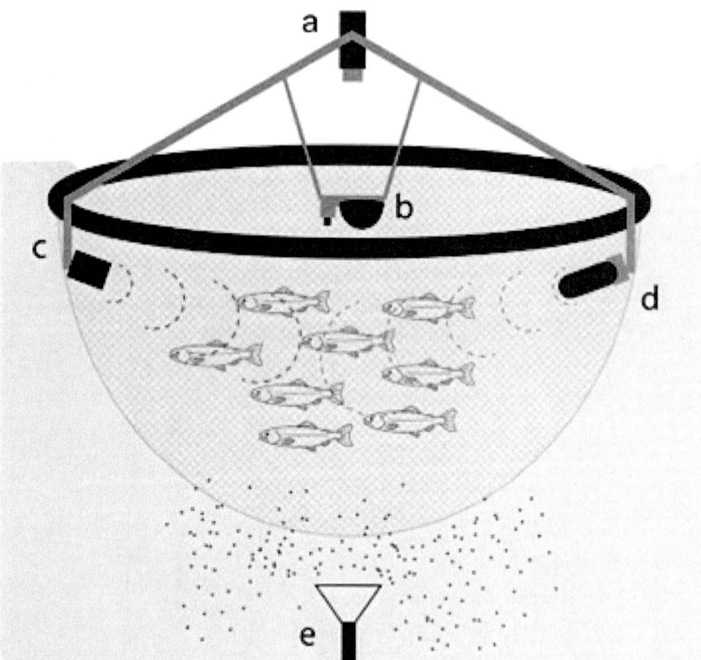

Fig. 1. Example of precision fish farming system including (a) surface camera, (b) underwater camera with multiparameter water probe, (c) sonar, (d) acoustic telemetry system with hydrophone, and (e) smart trap (Braña *et al.*, 2021).

2. Components of Precision Fisheries

Precision fisheries comprise several key components essential for effective fisheries management and sustainability:

2.1 Real-time monitoring

Real-time monitoring is a fundamental component of precision fisheries, providing fishermen with timely and accurate information to make informed decisions about their fishing operations. This monitoring system utilizes a variety of technologies, including sensors, satellite technology, and other monitoring devices, to collect data on various aspects of the marine environment, fish populations, and fishing activities.

2.1.1 Sensors: Sensors are deployed in the marine environment to collect data on parameters such as water temperature, salinity, dissolved oxygen levels, and turbidity. These sensors may be fixed to buoys, platforms, or underwater vehicles, allowing continuous monitoring of environmental conditions in real-time. By monitoring these variables, fishermen can identify optimal fishing locations based on the preferences of target species and environmental suitability.

2.1.2 Satellite Technology: Satellite technology plays a crucial role in real-time monitoring by providing fishermen with a bird's-eye view of the ocean. Satellites equipped with remote sensing instruments can detect various features of the marine environment, including sea surface temperature, chlorophyll concentration, ocean currents, and sea ice extent. Satellite imagery can also be used to monitor vessel movements, identify fishing hotspots, and detect illegal, unreported, and unregulated (IUU) fishing activities.

2.1.3 Monitoring Devices: In addition to sensors and satellites, monitoring devices such as acoustic tags, electronic monitoring systems, and vessel tracking systems are used to track fish movements, monitor fishing activities, and ensure compliance with fishing regulations. Acoustic tags are attached to individual fish to track their movements and behavior, providing valuable insights into fish migration patterns, habitat preferences, and stock dynamics. Electronic monitoring systems, including cameras and sensors installed onboard fishing vessels, record fishing activities, catch composition, and discards in real-time, facilitating accurate data collection and reporting. Vessel tracking systems, such as Automatic Identification Systems (AIS) and Vessel Monitoring Systems (VMS), enable authorities to monitor vessel movements, detect suspicious behavior, and enforce fisheries regulations.

By integrating data from these monitoring technologies, fishermen can gain a comprehensive understanding of the marine environment, fish populations, and fishing activities in real-time. This information allows them to make strategic decisions about when, where, and how to fish, optimizing their catch efficiency while minimizing environmental impact and compliance risks.

3. Predictive modeling

Predictive modeling in precision fisheries involves the application of sophisticated statistical and computational techniques to anticipate future trends in fish abundance, distribution, and behavior. These models draw upon a diverse array of data sources, including historical catch data, environmental variables such as sea surface temperature and chlorophyll concentration, oceanographic data, and biological parameters. By analyzing these datasets using advanced statistical algorithms, predictive models can forecast changes in fish populations over time and space.

Species distribution modeling (SDM) is a commonly used technique in precision fisheries, which predicts the spatial distribution of fish species based on habitat preferences and environmental conditions (Elith & Leathwick, 2009). SDM algorithms, such as MaxEnt and Generalized Additive Models (GAMs), utilize presence-absence or abundance data of fish species along with environmental variables to generate distribution maps that highlight areas with suitable habitat conditions for target species.

Catch forecasting is another essential aspect of predictive modeling in fisheries, aiming to estimate future fish abundance and catch potential in specific fishing grounds or regions. Time- series analysis, regression modeling, and machine learning algorithms are commonly employed to analyze historical catch data and environmental variables, enabling the prediction of future catch trends (Schnute & Richards, 2001).

Ecosystem modeling represents a more holistic approach to predictive modeling in fisheries, as it considers the complex interactions between fish populations, marine habitats, and environmental factors. These models, such as Ecopath with Ecosim (EwE) and Atlantis, simulate the dynamics of marine ecosystems by incorporating biological, physical, and socio- economic components (Fulton *et al.*, 2004). By simulating the effects of fishing pressure, environmental variability, and management interventions on ecosystem structure and function, ecosystem models provide valuable insights into the potential impacts of different management strategies on fish stocks and ecosystem health.

4. Targeted harvesting

Targeted harvesting in precision fisheries involves the implementation of selective fishing gear, real-time quota management systems, and spatial management measures to enable fishermen to selectively harvest specific species or size classes while minimizing bycatch and environmental impact (González-Troncoso *et al.*, 2020). Selective fishing gear, such as size and species-specific nets, traps, and hooks, allows fishermen to target their catch more precisely, reducing the capture of non-target species and undersized individuals. Real-time quota management systems utilize electronic monitoring and reporting technologies to track fishing activity and enforce catch limits in near real-time, ensuring compliance with fisheries regulations and preventing overfishing. Spatial management measures, including marine protected areas, fishing closures, and gear restrictions, designate areas where fishing activities are regulated or prohibited to protect sensitive habitats and species. By integrating these targeted harvesting techniques into fishing operations, precision fisheries can optimize catch efficiency while minimizing environmental impact, contributing to the long-term sustainability of marine ecosystems and fisheries resources.

5. Internet of Things (IoT)

In precision fisheries, the integration of Internet of Things (IoT) devices and sensors plays a pivotal role in revolutionizing fishing practices by enabling real-time monitoring of fishing activities, tracking fish movements, and collecting environmental data. IoT devices, such as underwater sensors, GPS trackers, and onboard cameras, are deployed in fishing gear, vessels, and fishing infrastructure to gather a wealth of data that can be used to inform decision-making and optimize fishing operations.

These IoT devices provide fishermen with valuable insights into various aspects of fishing, including fish abundance, distribution patterns, environmental conditions, and vessel performance. For example, underwater sensors can measure parameters such as water temperature, salinity, and dissolved oxygen levels, helping fishermen identify optimal fishing locations and assess habitat suitability for target species (Su *et al.*, 2018). GPS trackers installed on fishing vessels enable fishermen to monitor vessel movements in real-time, track fishing routes, and analyze fishing effort and catch patterns (Sturludóttir *et al.*, 2016). Onboard cameras capture video footage of fishing activities, allowing fishermen to observe catch composition, discards, and bycatch, and assess the effectiveness of fishing gear and techniques (Morato *et al.*, 2018).

6. Machine Learning and Artificial Intelligence (AI)

Machine learning and AI techniques are increasingly being applied in precision fisheries to analyze large datasets, detect patterns, and generate insights for fisheries management and decision support. These techniques allow for the development of predictive models that can forecast changes in fish populations, identify fishing hotspots, and optimize harvesting strategies. For example, machine learning algorithms can analyze historical catch data, environmental variables, and other relevant factors to predict future fish abundance and distribution patterns. AI-powered image recognition systems can also be used to identify fish species, estimate fish size, and assess catch composition from onboard camera footage. By leveraging machine learning and AI, precision fisheries can enhance decision-making, improve resource management, and promote sustainable fishing practices.

7. Fisheries Decision Support Systems

Fisheries decision support systems (DSS) are software tools and applications designed to assist fishermen, resource managers, and policymakers in making informed decisions about fishing operations, conservation measures, and policy interventions. These systems integrate data from various sources, including real-time monitoring data, environmental data, and fisheries management information, to provide users with actionable insights and recommendations. For example, DSS can be used to assess the impact of different management scenarios on fish populations, evaluate the effectiveness of conservation measures, and prioritize management actions based on their potential benefits and costs. By facilitating evidence-based decision- making and stakeholder engagement, fisheries DSS contribute to improved fisheries management, sustainability, and governance.

8. Collaborative Research and Stakeholder Engagement

Collaborative research and stakeholder engagement play a crucial role in the development and implementation of precision fisheries initiatives. By bringing together fishermen, scientists, policymakers, and other stakeholders, collaborative research efforts facilitate the co-design, co-implementation, and co-evaluation of precision fisheries projects. Stakeholder engagement processes ensure that the needs, priorities, and perspectives of all relevant stakeholders are taken into account, enhancing the relevance, effectiveness, and acceptance of precision fisheries initiatives within the fishing community and beyond. Collaborative research and stakeholder engagement also foster knowledge sharing, capacity building, and mutual learning, leading to more inclusive and participatory decision-making processes in fisheries management and governance.

9. Challenges and Opportunities

While precision fisheries hold immense promise, several challenges must be addressed:

- Technological Barriers: High costs, limited infrastructure, and lack of technical expertise hinder the adoption of advanced technologies and data-driven approaches.

- Data Accessibility and Quality: Timely and reliable data on fish stocks, environmental conditions, and fishing activities are essential for effective decision-making and management.
- Social and Institutional Factors: Stakeholder collaboration, regulatory frameworks, and governance structures play a crucial role in the successful implementation of precision fisheries initiatives.

Conclusion

Precision fisheries represent a paradigm shift in fishing practices, offering innovative solutions to address the complex challenges facing fisheries management and sustainability. By integrating advanced technologies, data-driven approaches, and ecosystem-based principles, precision fisheries enable fishermen to optimize operations, minimize environmental impact, and ensure the long-term viability of marine ecosystems and coastal communities.

References

Benoît, H. P., Outeiro, L., Teles-Machado, A., & Teodósio, M. A. (2019). Fish Image Recognition in Marine Environmental Monitoring-A Review. Sensors, 19 (19), 4238.

Braña, C. B. C., Cerbule, K., Senff, P., & Stolz, I. K. (2021). Towards Environmental Sustainability in Marine Finfish Aquaculture. Frontiers in Marine Science, 8.

Elith, J., & Leathwick, J. R. (2009). Species distribution models: ecological explanation and prediction across space and time. Annual Review of Ecology, Evolution, and Systematics, 40, 677-697.

Fulton, E. A., Smith, A. D. M., Smith, D. C., & Van Putten, I. E. (2014). Human behaviour: the key source of uncertainty in fisheries management. Fish and Fisheries, 15(3), 427-449.

García-Seoane, J. J., Lourenço, P., & Jiménez-Valverde, A. (2020). Machine Learning Models for Predicting Fish Distribution and Abundance in Marine Ecosystems: A Review. Frontiers in Marine Science, 7, 583.

González-Troncoso, D., Villasante, S., da Rocha, J. M., & López-López, L. (2020). Influence of fishing gear selectivity and effort control on the exploitation patterns of the European hake (Merluccius merluccius L.). ICES Journal of Marine Science,77 (3), 1080-1090.

Morato, T., Pham, C. K., & Pinto, C. (2018). A perspective on fisheries-related observations using underwater cameras—A review. Fisheries Research, 204, 199-210.

Plagányi, É. E., van Putten, I., Hutton, T., & Hutton, T. (2014). Integrating indigenous livelihood and lifestyle objectives in managing a natural resource. Proceedings of the National Academy of Sciences, 111 (12), 4528-4533.

Schnute, J. T., & Richards, L. J. (2001). Forecasting recruitment using the retrospective patterns of spawner-recruit systems. Canadian Journal of Fisheries and Aquatic Sciences, 58 (2), 393-404.

Sturludóttir, E., Scheving, L. A., Andersen, K. H., & Sigurdsson, T. (2016). Real-time vessel monitoring data in fisheries management: applications and challenges. ICES Journal of Marine Science, 73(9), 2214-2222.

Su, G., Zhang, H., & Xu, C. (2018). Real-time monitoring and prediction of fishing vessel behaviors based on data streams. Ocean Engineering, 156, 206-215.

9

Sustainable Aquaculture Practices and Approaches

Kamin Alexander*[*] *and Shruti Samson

Department of Biological Sciences, Sam Higginbottom University of Agriculture Technology and Sciences, Prayagraj-211007, Uttar Pradesh, India

Abstract

Sustainable aquaculture is crucial for meeting the increasing global demand for seafood while minimizing environmental impact and ensuring long-term viability. This chapter explores various sustainable aquaculture practices and approaches aimed at enhancing productivity, environmental stewardship, and socio-economic benefits. From integrated multitrophic aquaculture to recirculating aquaculture systems, this chapter examines innovative techniques and management strategies that promote sustainability across the aquaculture sector. Through case studies and best practices, the chapter highlights the importance of adopting holistic approaches to aquaculture that consider ecological, social, and economic dimensions.

1. Introduction

Aquaculture, the cultivation of aquatic organisms, has emerged as a critical component of global food production, contributing significantly to meeting the increasing demand for seafood. Over the past few decades, aquaculture has experienced exponential growth, fueled by advancements in technology, increasing consumer demand, and declining wild fish stocks. This expansion has positioned aquaculture as one of the fastest-growing food sectors worldwide, providing essential protein sources and livelihoods for millions of people (Ayyappan *et al.*, 2006).

Despite its undeniable contributions to food security and economic development, the rapid expansion of aquaculture has raised significant concerns regarding its environmental and social sustainability. The intensification of aquaculture operations has led to a range of environmental challenges, including habitat degradation, pollution of water bodies, and the introduction of invasive species. Moreover, the concentration of aquaculture activities in certain regions has exacerbated social issues such as displacement of coastal communities, conflicts over resource use, and labor rights violations.

In response to these challenges, there has been a growing recognition of the need for sustainable aquaculture practices and approaches that prioritize environmental stewardship, social equity, and economic viability. Sustainable aquaculture aims to strike a balance between meeting the growing demand for seafood and safeguarding the health of aquatic ecosystems and the well- being of local communities. It encompasses a range of principles and strategies aimed at minimizing environmental impact, conserving natural resources, and promoting responsible management of aquatic resources (Pauly *et al.,* 2002).

Key objectives of sustainable aquaculture include reducing reliance on wild fish stocks through responsible sourcing of feed ingredients, minimizing the use of antibiotics and chemicals, optimizing water and energy use, and enhancing biodiversity conservation. Furthermore, sustainable aquaculture seeks to foster social inclusivity, empowering local communities to participate in decision-making processes, access resources, and share benefits from aquaculture development.

This chapter explores the multifaceted dimensions of sustainable aquaculture practices and approaches, highlighting innovative techniques, case studies, and best practices from around the world. By examining the principles of integrated multitrophic aquaculture, recirculating aquaculture systems, organic aquaculture, and community-based aquaculture, we aim to elucidate the diverse strategies employed to promote sustainability and resilience in the aquaculture sector. Through collaborative efforts and continuous improvement, sustainable aquaculture holds the promise of providing nutritious food, supporting livelihoods, and safeguarding the health of aquatic ecosystems for generations to come.

2. Integrated Multitrophic Aquaculture (IMTA)

Integrated Multitrophic Aquaculture (IMTA) represents a pioneering approach in sustainable aquaculture, emphasizing the cultivation of multiple species within a single ecosystem to maximize resource utilization and minimize environmental impact. In IMTA systems, species from different trophic levels, including finfish, shellfish, and seaweeds, are strategically integrated to create symbiotic relationships that enhance overall system productivity and resilience.

The cornerstone of IMTA lies in the principle of nutrient cycling, where waste products from one species serve as inputs for another, thereby promoting efficient resource utilization and reducing environmental pollution. For example, fish excretions, uneaten feed, and organic matter accumulate in the water column, providing nutrients for the growth of seaweeds and bivalves (Chopin *et al.*, 2008). Seaweeds, in turn, absorb excess nutrients and mitigate eutrophication, while bivalves filter particulate matter and enhance water quality. This integrated approach mimics natural ecosystems, where nutrient flows are efficiently recycled and redistributed, contributing to overall system stability and productivity.

Successful IMTA projects have been implemented worldwide, showcasing the potential of this approach to improve sustainability and resilience in aquaculture. For instance, salmon-seaweed IMTA systems have been established in Norway, where Atlantic salmon are cultured alongside kelp and other seaweeds. These systems not only reduce the environmental impact of salmon farming by utilizing excess nutrients but also diversify revenue streams for farmers through the sale of seaweed products.

Similarly, shrimp-bivalve IMTA systems have been developed in Southeast Asia, where shrimp ponds are integrated with mangrove forests and shellfish beds (Chopin *et al.*, 2001). Mangroves provide habitat and refuge for juvenile shrimp, while shellfish filter pond effluents and enhance water quality (Chopin *et al.*, 2001). These integrated systems improve the ecological resilience of shrimp farming operations, reduce disease outbreaks, and enhance biodiversity conservation in coastal areas.

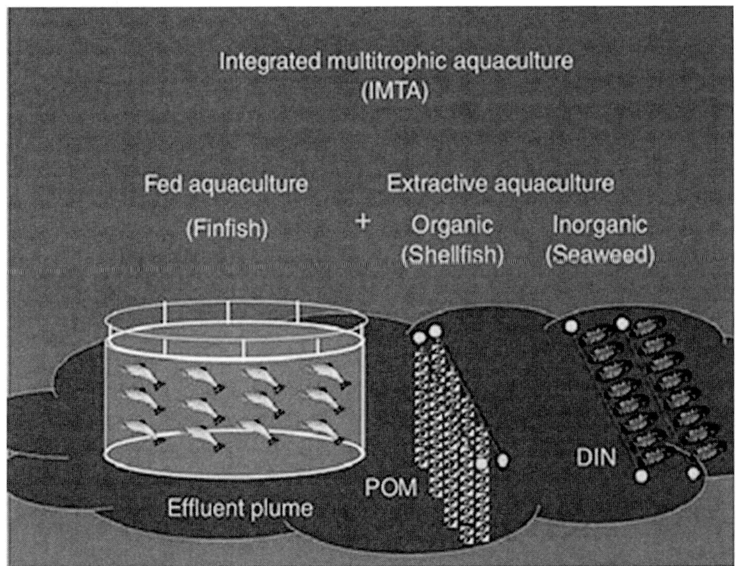

Fig. 1. Conceptual diagram of an integrated multitrophic aquaculture (IMTA) (Chopin *et al.*, 2008)

In addition to environmental benefits, IMTA offers economic advantages by diversifying revenue streams, reducing input costs, and increasing overall system productivity. By harnessing synergies between different species and optimizing ecosystem functions, IMTA holds great promise for sustainable aquaculture development in the face of growing environmental and socio-economic challenges.

3. Synergistic Interactions

The key principle underlying IMTA is the promotion of synergistic interactions between different species to enhance overall system productivity and resilience. For example, fish excretions and uneaten feed provide nutrients for the growth of seaweeds, while seaweeds absorb excess nutrients and mitigate eutrophication. Similarly, shellfish filter particulate matter and enhance water quality, creating a conducive environment for fish growth and survival.

3.1 Case Studies and Success Stories

Numerous successful IMTA projects have been implemented worldwide, demonstrating the feasibility and effectiveness of this approach in enhancing sustainability and resilience in aquaculture. For example, the Integrated Salmon Aquaculture (ISA) project in Norway integrates Atlantic salmon farming with seaweed cultivation and shellfish farming, leading to improved water quality, reduced environmental impact, and enhanced economic viability.

4. Recirculating Aquaculture Systems (RAS)

Recirculating aquaculture systems (RAS) are closed-loop aquaculture systems that recycle and treat water within a controlled environment, minimizing water usage and discharge. RAS technology allows for intensive fish production in land-based facilities while reducing environmental impact, including water pollution and habitat degradation. By optimizing water quality, temperature, and nutrient levels, RAS systems provide a stable and controlled environment for fish growth, leading to higher yields and improved biosecurity. Case studies of RAS facilities, such as land-based salmon farms and indoor shrimp production facilities, illustrate the potential of RAS to increase efficiency and sustainability in aquaculture.

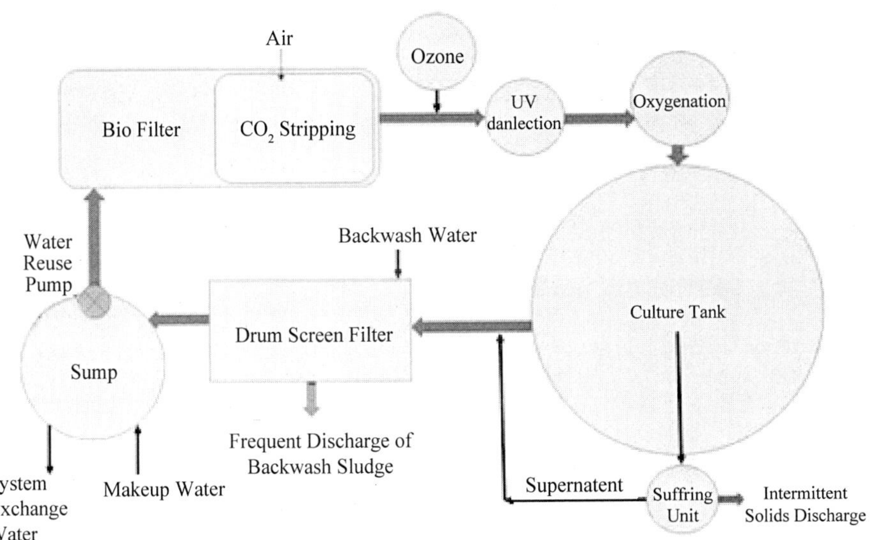

5. Community-Based Aquaculture

Community-based aquaculture initiatives involve local communities in the planning, management, and benefits of aquaculture activities, promoting social equity, and economic development. These initiatives empower communities to participate in decision-making processes, access resources, and share benefits from aquaculture development. By fostering collaboration, knowledge exchange, and capacity building, community-based aquaculture contributes to poverty alleviation, food security, and sustainable livelihoods. Case studies of community-based aquaculture projects, such as small-scale fish farming cooperatives and community-managed aquaculture zones, highlight the importance of social inclusion and community engagement in aquaculture development.

Conclusion

Sustainable aquaculture practices and approaches are essential for addressing the environmental, social, and economic challenges facing the aquaculture sector. By adopting holistic and innovative techniques such as integrated multitrophic aquaculture, recirculating aquaculture systems, organic aquaculture, and community-based aquaculture, the industry can promote sustainability, resilience, and prosperity. Through collaboration, knowledge sharing, and continuous improvement, sustainable aquaculture can contribute to food security, environmental conservation, and economic development worldwide.

References

Ayyappan, S & Diwan, Arvind. (2006). Fisheries research and development in India. 26.

Chopin, T., Buschmann, A. H., Halling, C., & Yarish, C. (2001). Integrating seaweeds into marine aquaculture systems: a key toward sustainability. Journal of Phycology, 37(6), 975–986. https://doi.org/10.1046/j.1529-8817.2001.01137.x

Chopin, T., Robinson, S., Troell, M., Neori, A., Buschmann, A., & Fang, J. (2008). Multitrophic Integration for Sustainable Marine Aquaculture. In Elsevier eBooks (pp. 2463–2475).

Holan, A. B., Good, C., & Powell, M. D. (2020). Health management in recirculating aquaculture systems (RAS). In Elsevier eBooks (pp. 281–318).

Neori, A., Troell, M., Buschmann, & Yarish, C. (2004). Integrated aquaculture: rationale, evolution and state of the art emphasizing seaweed biofiltration in modern mariculture. Aquaculture, 231(1–4), 361–391. https://doi.org/10.1016/j.aquaculture.2003.11.015

Pauly, D., Christensen, V., Guénette, S., Pitcher, T. J., Sumaila, U. R., Walters, C. J., Watson, R., & Zeller, D. (2002). Towards sustainability in world fisheries. Nature, 418(6898), 689–695. https://doi.org/10.1038/nature01017

10

Recent Trends and Innovative Technology in Fishing Technology

Prabhutva Chaturvedi*, Binal Rajeshbhai Khalasi, Mayurkumar U. Tandel, Bhavy Dalsaniya and Kirankumar Gopalbhai Baraiya

College of Fisheries, Mangaluru, Karnataka Veterinary, Animal and Fisheries Sciences University, Bidar-575002, Karnataka, India

Abstract

The major concern of fisheries and aquaculture is to reduce the widening gap between fish demand and supply and achieve the ultimate goal of self-sufficiency in fish production; however, this can only be achieved if technology advances. The world's fishery resources require scientific management to preserve their long-term viability and availability for future generations. Overfishing and its detrimental effects on ecosystems have become increasingly evident in recent years. Recent developments in fishing technology have focused on reducing the bycatch of non-target species, protected species, and juveniles, minimizing the environmental impact of fishing gear and their operation, and minimizing the energy use per unit volume of fish landed during fishing operations. Fishing innovations have been around for thousands of years, and they were primarily motivated by the goal of enhancing capture efficiency, which was sparked by a growing comprehension of the habits and behaviors of the species that were being sold. The degree of adoption of responsible fishing technologies strongly depends on the robustness of the fisheries management system.

Keywords: *Fishing technologies, Overfishing, bycatch, Innovation, fisheries management*

1. Introduction

The increase in the consumption of fish in the human diet is caused by a shortage of animal-based proteins. The demand for fish has grown over time, but the supply has never kept up with the demand. However, the main concerns of fisheries and aquaculture have been narrowing the growing gap between fish supply and demand and reaching the ultimate objective of self-sufficiency in fish production. High-sea fishing is now feasible owing to advancements in refrigeration technology (Wang and Wang 2005), vessel technology, echo-sounding and navigational equipment, and other technical developments in fishing gear (Johnsen 2005; Glass *et al.*, 2007; Tidd *et al.*, 2017). Generally, it refers to an organization that raises or harvests fish that are recognized as a fishery by some body (Fletcher *et al.*, 2002).

"Innovation" in the context of fisheries refers to the creation and use of fresh concepts, methods, technologies, and strategies that enhance the industry's overall performance, sustainability, and efficiency. Innovation in fisheries, according to the International Council for the Exploration of the Sea (ICES), is an improvement over the current state of affairs, whether incremental, transformational, or descriptive (WKING; ICES 2020). Fishing innovations have been around for thousands of years, and they were primarily motivated by the goal of enhancing capture efficiency, which was sparked by a growing comprehension of the habits and behaviors of the species that were being sold. Technological innovations such as steam-engine vessels, onboard refrigeration, freezing of catches, synthetic netting materials, and information technologies to help with communication, navigation, fish location, and gear performance monitoring during fishing were brought about by the industrialization of fisheries in the 19th and 20th centuries, and have contributed to modern times (Squires and Vestergaard, 2013). These developments have led to the development of much larger fishing boats and gear, which has made it possible to exploit fish populations at much greater productivity levels and in previously unreachable oceanic areas and depths. The expansion and development of fisheries were facilitated by this unchecked technological advancement, but it also paved the way for overexploitation and the ensuing negative environmental effects. Driven by management requirements and changing consumer preferences for “sustainable” seafood, recent innovations in fisheries have turned toward promoting sustainable practices, reducing bycatch, and minimizing environmental impacts (Lewison *et al.*, 2014; Kennelly and Broadhurst, 2021; Hilborn *et al.*, 2023).

The development and implementation of sophisticated fishing technology have been driven by the imperatives of minimizing the impact on non-target species, reducing bycatch, and adapting fishing tactics to changing ocean conditions owing to climate change (Cheung *et al.*, 2013). Furthermore, economic considerations, such as growing fuel prices and the requirement for greater operational effectiveness, function as strong catalysts for the ongoing development of fishing vessel design and navigation systems (Kasperski and Holland, 2013). Innovations in data collection and management technology have resulted from customer demand for sustainable seafood and the desire to improve transparency and traceability in seafood supply chains. The impact of trawl fisheries on benthic habitats and ecosystems, particularly in areas with complex biogenic structures, has drawn special attention (Watling and Norse, 1998), and public opinion is moving in favor of switching to less damaging, if frequently less effective, fishing gears. Nonetheless, recent studies suggest that through technological innovation and careful management, all gear types may be fished sustainably (Hilborn *et al.*, 2023).

2. Recent developments in fishing technology

The technology used to collect fish has seen rapid improvements over the past several decades. Among the most significant developments that have affected the historical evolution of fishing gear and practices have been (i) developments in craft technology and mechanization of propulsion, gear, and catch handling; (ii) introduction of synthetic gear materials; (iii) developments in acoustic fish detection and satellite-based remote sensing techniques; (iv) advances in electronic navigation and communication equipment; and (v) awareness of the need for responsible fishing to ensure sustainability of the resources, protection of biodiversity, environmental safety, and energy efficiency. The introduction of strong and extremely effective fish harvesting systems and fish identification techniques, along with an unchecked fleet size increase driven by the growing market demand for fish, has resulted in a rising strain in the world's fisheries resources. The world's fishery resources require scientific management to preserve their long-term viability and availability for future generations. Overfishing and its detrimental effects on ecosystems have become increasingly evident in recent years.

3. The major technological changes that have taken place in India's capture fisheries are

- Creation and spread of synthetic materials used in fishing gear.
- The use of automated purse seining and trawling
- Increase in the size, quantity, installed hand capacity, and introduction of multi-day fishing in the automated fleet.
- Increased productivity and variety in trawls, purse seines, gillnets, and lines for the mechanized industry.
- Expansion of fishing areas for catching lobsters, cephalopods, and deep-sea prawns.
- Adoption of current technology, such as echo sounders and GPS.
- Motorization of traditional fishing vessels and growth of fishing grounds.
- Improvement of classic fishing units in terms of craft modernization, gear materials, gear efficiency, and proportions.

4. Fishing Vessels

Over the course of the 20th century, fishing vessel designs and equipment have undergone significant changes in tandem with the industry's growth, heightened competition for scarce resources, and the necessity of venturing farther to locate fish to procure hydraulic or electrical equipment for handling gear and different methods for preserving the catch.

The ability of vessels to handle, process, and store fish in excellent conditions onboard has significantly expanded with the introduction of ice and refrigeration technology. There have been significant advancements in fishing vessels in terms of propulsion systems, hull optimization, energy conservation, pre-processing, processing, preservation, packaging systems, gear and catch handling deck equipment, enhanced navigational equipment, fish-finding electronics, and craft materials. Small-scale fishing operations in developing nations have seen significant changes as a result of the invention and widespread use of outboard engines. The command console of a large contemporary fishing vessel is similar to that of an aircraft cockpit. Navigation controls and primary displays are displayed on monitors in front of the control position; these are increasingly displayed on a single, large integrated display. Instruments include radio communications that are crucial for both general and safe communications, as well as fish detection and fishing operation tools such as sonar and electronic assistance. Navigational instruments are used for vessel navigation, both at sea and in harbors.

The Global Maritime Distress and Safety System (GMDSS) and satellite-based vessel position monitoring programs have improved the safety of larger fishing vessels; however, no international legal instrument specifically addressing fishing vessel safety has yet been put in place. Owing to a lack of proper regulations and enforcement, the safety of ships and their personnel is frequently ignored in developing nations.

5. Fishing Gear Materials

Traditional fishing gear in the past was less productive and less effective. Natural fibers, including cotton, manila, sisal, jute, and coir, were used to make them. These materials have limited service life and significant maintenance costs because they are susceptible to biodegradation. Along with the development of other contemporary materials, fiber technology has undergone significant advancements in recent decades. Natural fibers used for fishing gear were replaced by synthetic materials when they were introduced in India in the late 1950s. This was because the synthetic materials had higher breaking strengths, greater resistance to weathering, lower maintenance costs, longer service lives, and more consistent characteristics. Synthetic fibers such as polyamide (PA), polyester (PES), polyethylene (PE), and polypropylene (PP) are most commonly used in fishing. Other synthetic fibers that are less frequently employed and typically confined to Japanese fisheries include polyvinyl alcohol (PVAA), polyvinyl chloride (PVC) and polyvinylidene chloride (PVD). Ultrahigh molecular weight polyethylene (UHMWPE), often known as "dyneema," is 15 times stronger than steel and is one of the newest, higher-tensile-strength materials that have been launched recently. The fishing net design and size have changed as a result of the advent of synthetic materials with high tensile strength. Netting was previously performed by hand, which is time-consuming and tedious. Currently, practically all fishing nets are constructed using machine-made nettings.

6. Fishing Gears

Fishing gear can be regarded as any form of equipment utilized in harvesting, farming, or collecting fish from any water body (Nuhu and Yaro, 2005). Whether primitive or sophisticated, fishing gear uses five mechanisms to capture fish: gilling and tangling (using tools such as harpoons and gill nets), trapping (using tools such as traps and pound nets), filtering (using tools such as trawls, seines, and other net fishing systems), hooking and spearing (using tools such as hooks and lines and harpoons), and pumping (using tools such as fish pumps).

The concepts of catch, design, technological characteristics, and operating procedures were used to categorize fishing gear. The major categories of fishing gear are defined and categorized by the FAO as follows.

- Surrounding nets (including purse seines)
- Seine nets (including beach seines and boat, scottish/danish seines)
- Trawl nets (including bottom: beam, otter and pair trawls, and midwatertrawls: otter and pair trawls)
- Dredges
- Lift nets
- Falling gears (including cast nets)
- Gillnets and entangling nets (including set and drifting gillnets; trammelnets)
- Traps (including pots, stow or bag nets, fixed traps)
- Hooks and lines (including handlines, pole and lines, set or driftinglonglines, trolling lines)
- Grappling and wounding gears (including harpoons, spears, arrows,etc.)
- Stupefying devices

Historically, the goal of technological advancements in fishing gear and techniques has been to boost productivity by increasing gear system efficiency. However, with overfishing currently occurring and more people being aware of the negative effects fishing has on the environment and ecosystems, the development of fishing gear is concentrated on creating systems that are responsible, have better species- and size-specific qualities, have fewer adverse effects on the environment and non-target resources, and maintain sustainable fish populations. Purse seines and trawls rank first and second in importance, respectively, among the wide range of harvesting techniques used in commercial fisheries worldwide, behind gill nets, lines, and entangling nets and traps.

However, in line with technological advancements, fishing gear has evolved significantly in recent years, despite the fact that the fundamental principles of hurting, hooking, capturing (gills, fins, and spines), surrounding, scooping, and filtering are still present. The advancement of fishing vessels coincided with that of fishing gear. Since fishing had to be done beyond the banks and shorelines, a boat or vessel that could help fishermen enter deeper water was required. Most of these antiquated trades are practiced today, especially in emerging tropical nations. However, as with fishing gear, modern fishing vessels have largely changed in terms of size, quality, and sophistication, all of which are made possible by technological advancements. The advancement of fish detection techniques and the creation of new fish finding and detection equipment, such as echo sounders and sonars, came after the development of sophisticated fishing gear and the improvement of vessel designs and sizes.

7. Technologies for Responsible fishing

The technologies available for responsible fishing are centered on minimizing the environmental effects of fishing gear and its operation, limiting the bycatch of non-target species, juveniles, and protected species, and minimizing the energy used per unit volume of fish landed during fishing operations.

8. Bycatch reduction in fishing

Whatever non-targeted animal is kept, sold, or thrown away for whatever reason is referred to as bycatch. "Target catch" refers to the species or species assemblage that is primarily sought in a fishery (such as shrimp and cephalopods); "incidental catch" is the retained catch of non-targeted species; and "discarded catch" is the portion of catch that is returned to the sea due to economic, legal, or personal considerations. The catch process and production of bycatch during trawling are represented in long-lived species with sluggish population growth rates, such as seabirds, marine mammals, and sea turtles, which are particularly vulnerable to bycatch (Heppell *et al.* 2005). Despite the development of more selective fishing methods, bycatch (i.e., creatures that are inadvertently taken) remains an important challenge for fisheries management, policy, and science (FAO 2011; Gilman *et al.*, 2020).

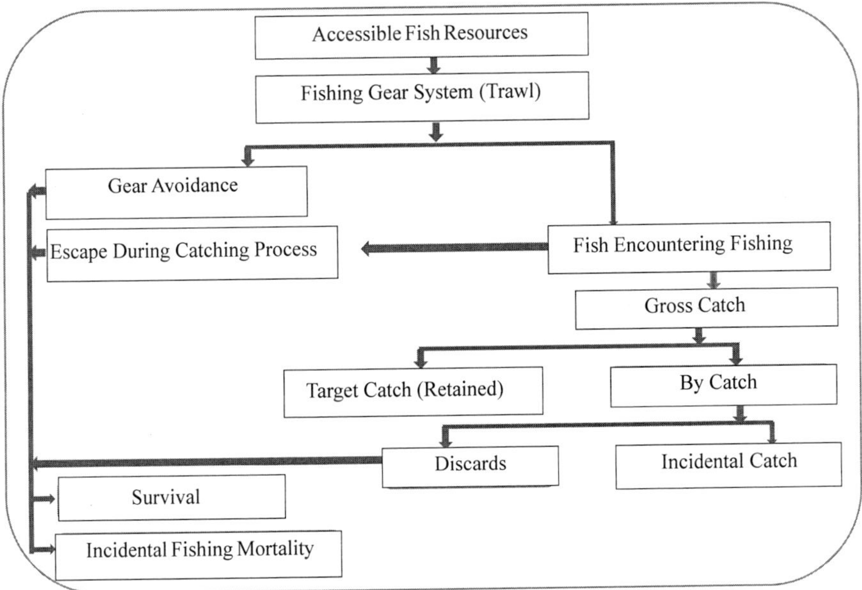

Fig. 1. Catch process and bycatch production in trawling

Fishermen have acknowledged the significance of limiting the ecological effects of fishing operations and decreasing bycatch, as have scientists and fisheries management. The development of selective fishing gear is emphasized in the Code of Conduct for Responsible Fisheries (FAO, 1995; 1996) as a means of conserving resources, safeguarding non-targeted resources, and preserving endangered species, such as sea turtles.

9. Bycatch Reduction Devices

Bycatch Reduction Devices are designed to decrease the number of non-targeted species captured during trawling (BRDs). These devices were created based on the understanding that shrimp and other creatures inside the net vary in size and exhibit various behavioral patterns. Various bycatch reduction devices have been developed in the fishing industry worldwide (Boopendranath *et al.*, 2008; 2009).

Based on the type of material utilized in their construction, BRDs can be roughly divided into three categories: (i) Soft BRDs, (ii) Hard BRDs, and (iii) Combination BRDs. To separate and exclude bycatch, soft BRDs employ soft materials such as rope frames and netting. Hard BRDs separate and exclude bycatch using rigid or semirigid grids and structures. Combination BRDs combine multiple BRDs into a single system and often combine hard and soft BRDs.

A popular method to reduce bycatch in shrimp trawls is the use of BRDs. Some of the advantages of reducing the amount of unwanted bycatch caught in shrimp trawls using BRDs are (i) reduction in the impact of trawling on non-targeted marine resources, (ii) reduction in damage to shrimp due to the absence of large animals in codends, (iii) shorter sorting times, (iv) longer tow times, and (v) lower fuel costs due to reduced net drag (Boopendranath & Pravin, 2009). According to reports, the installation of a BRD has little impact on the overall drag of the trawl system and thus on fuel consumption (Boopendranath *et al.*, 2008).

10. Approaches for reducing bycatch in major fishing gear systems

Fishing gear system	Gear design related approaches	Operation related approaches
Trawls	Trawl design; mesh size; bycatch reduction devices and turtle excluder devices; juvenile and trash fish excluder devices	Choice of fishing area, fishing depth, fishing time and season.
Purse seines	Seine design and seine depth appropriate for schools of target species; mesh size; excluder devices, aprons and Medina panel.	Choice of fishing area, fishing depth, fishing time and season; Capability of vessel and crew to use selective manoeuvres.
Gill nets	Mesh size; netting material; hanging ratio; use of scaring devices and acoustic deterrents	Choice of fishing area, fishing depth, fishing time and season.
Lines	Hook design, shape and size; hook spacing	Choice of bait type and bait size, fishing area, fishing depth, fishing time and fishing season; Use of dyed baits, side sets, subsurface line setting chutes and bird scaring steamers to deter birds; Use of circle hook and deep setting line to minimise sea turtle bycatch; Use of rare earth magnets to deter sharks.
Traps	Trap design; optimised trap mouth; escape windows	Choice of bait type, fishing area, fishing depth, fishing time and season.

10.1 Trawl Fisheries

Trawl trawling plays a significant role in the marine fishing industry in Southeast Asia. Millions of people live in coastal towns and rely on them for food and livelihoods. It also supplies feed to expand the aquaculture industry in this area. The trawl fishing industry has made an effort to compensate for the decline in profits by focusing more on vast quantities of low-value, small-sized fish, or "trash fish," as they are referred to in Southeast Asia (Funge-Smith *et al.* 2005; Nguyen 2017). Numerous issues plague trawl fisheries, such as overcapacity, overfishing, low profitability, and insufficient control. whatever non-targeted species are kept, sold, or thrown away for whatever reason is referred to as bycatch (Alverson *et al.*, 1994). In addition to the bycatch problem, the unmeasured effects of various fishing methods on the environment are significant. Fishing has a greater influence on ecology when using active fishing gear such as trawls. Because a large percentage of juveniles and subadults, especially economically valuable fish, make up bycatch in the tropics, bycatch reduction methods (BRD) require careful consideration and development. Given the significance of bycatch management and discard reductions in responsible fisheries, the FAO has released international recommendations on these topics (FAO, 2011). The Sustainable Development Goal (SDG) number fourteen, "Life under Water," incorporates many objectives for the sustainable use of fishery resources.

10.1.1 Square mesh window

A cod end with a square mesh window, as opposed to a full square mesh cod end or a traditional diamond mesh cod end, is frequently a more adaptable and useful way to keep small fish. Once within the codend, the fish often swim back through the holes in the front part of the codend to escape. The benefit of a square mesh is that the mesh hole is not deformed during the operation, unlike diamond meshes (Robins *et al.*, 1999, Boopendranath *et al.*, 2008). The trials also suggested that a traditional cod end fitted with a square-mesh window is a more flexible and practical way of rejecting undersized fish than a full square-mesh cod end. In experiments conducted by scientists to reduce the effect of trawl fishing on protected sea snakes, the square mesh codend also fared well. According to Courtney *et al.* (2008), square-mesh codend BRDs were proven to be extremely successful in reducing sea snake capture by 60%.

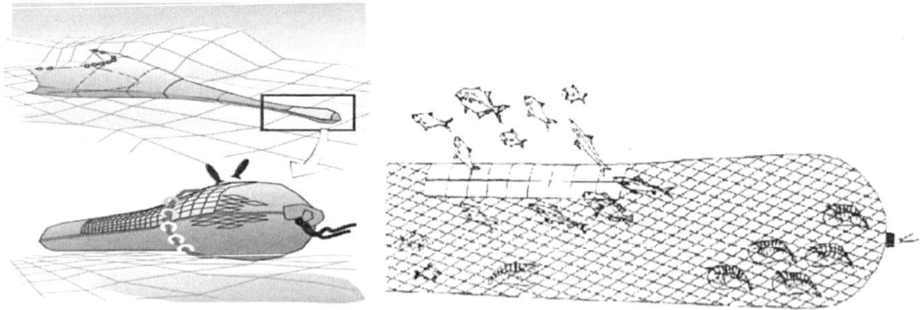

Fig. 2. Cod end equipped with a square mesh window

10.1.2 Bigeye BRD

Bigeye BRD is defined by a straightforward horizontal incision in the upper region of the codend or hind belly, which is bound with twine or kept open with floats and sinkers. Once within the codend, the fish often swim back through the holes in the front part of the codend to escape. The aperture (or horizontal slit) of the bigeye BRDs had the same proportions as the fisheye BRD. The goal of the current study was to evaluate the effectiveness of the bigeye bycatch reduction device and the best location inside the codend for use in automated shrimp trawls operating in the Indian seas.

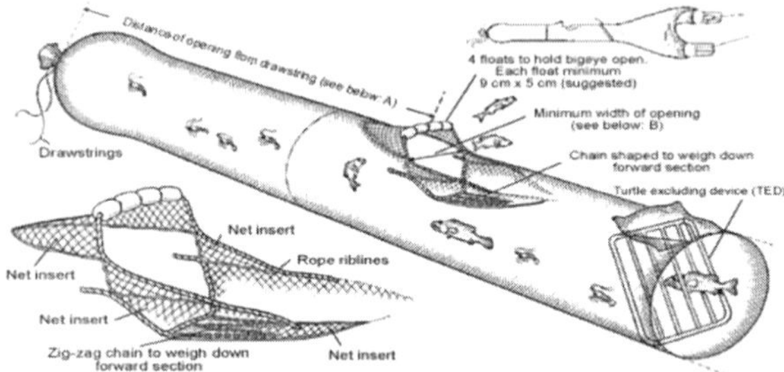

Fig. 3. Bigeye bycatch reduction device

10.1.3 Fisheye BRD

The fisheye BRD is a semi-round aperture that is fixed at a specific distance from the codend's drawstring by a stiff, triangular frame. This stiff mechanism helps fish escape the codend, particularly in smaller fish. Using a metal brace, the fisheye BRD supported a permanent escape opening in the top seam of the codend. Fish escape mostly occurs during haul back at the surface. The escape entrance of the fisheye must always be free from obstructions from additional net modifications or the lazy line connection mechanism for it to function properly in the water. There is not much further upkeep required for the sturdy metal frame. When fitted correctly, fisheye BRD has been reported to eliminate 37% of bycatch and retain 90% of shrimp by weight.

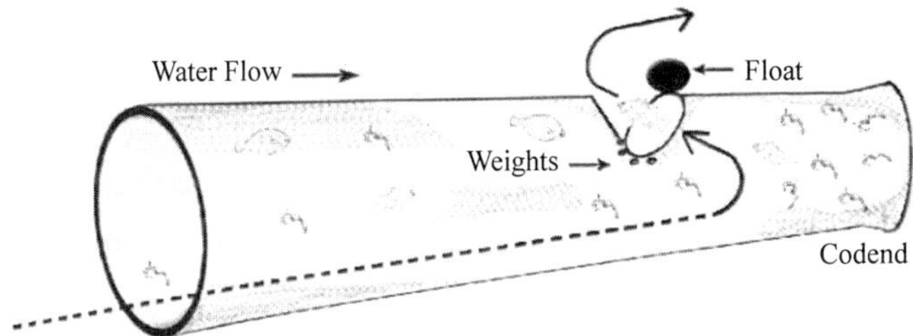

Fig. 4. Diagram of a typical fisheye excluder BRD

10.1.4 Rigid grid Brds

The separation of shrimp from non-shrimp resources has been accomplished through the development of several stiff grid sorting device designs, including the Nordmore grid (Isaksen *et al.*, 1992), juvenile fish excluder cum shrimp sorting device (JFE-SSD) (Boopendranath *et al.*, 2008), and juvenile and trash excluder device (JTED) (Chokesanguan *et al.*, 2000). In the Norwegian seas, Nordmore grid operations have demonstrated a low shrimp loss of 2–5%.

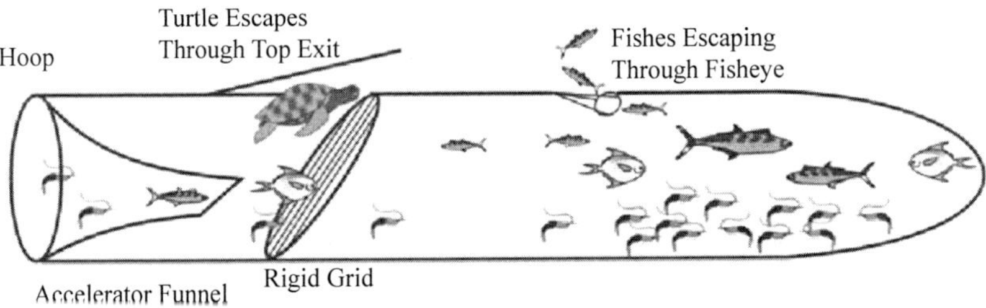

Fig. 5. Principle of operation of rigid grid sorting device

10.1.5 Sieve net Brd

According to Polet *et al.*, (2004), a sieve net is a large mesh funnel that is fixed inside the net and directs fish to a second codend with huge diamond mesh netting, whereas shrimp pass through large meshes and gather in the main codend. Sieve nets, also called veil nets, are cone-shaped nets that are placed inside regular trawls to drive undesired bycatch to an escape hole drilled into the trawl's body, leading to a second codend. Sieve nets are utilized in commercial shrimp fleets in the Netherlands, Denmark, the UK, France, Germany, and Belgium (Polet *et al.*, 2004; Revill and Holst, 2004). In sieve net-installed trawl operations at various fishing sites, bycatch exclusion rates of 15–50% have been observed along with shrimp losses of 5–15% (Polet *et al.*, 2004).

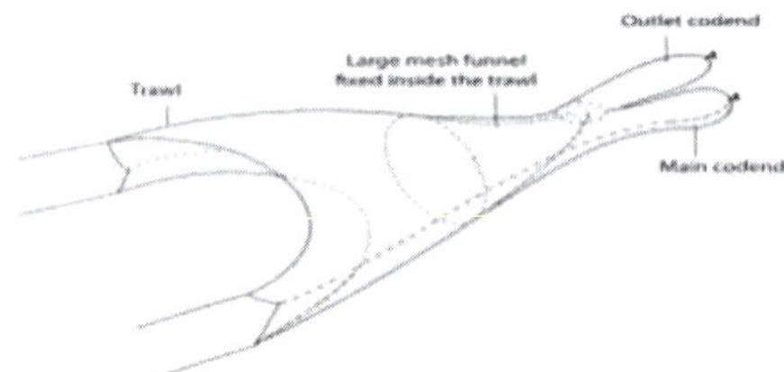

Fig. 6. Perspective view of Sieve net BRD installed in the trawl net

10.1.6 Juvenile Fish Excluder cum Shrimp Sorting Device (JFE-SSD)

The standard codend in a trawl is replaced with the Juvenile Fish Excluder cum Shrimp Sorting Device (JFE-SSD), which is a bycatch reduction device with an in situ sorting mechanism. CIFT has created a unique solution to this problem by producing a Juvenile Fish Excluder cum Shrimp Sorting Device (JFE-SSD) (Boopendranath *et al.*, 2008; WWF, 2009a). In commercial shrimp trawls, JFE-SSD reduces the bycatch of juveniles and small-sized non-targeted species while allowing fishermen to collect and maintain economically desirable finfish and shrimp species. Higher catch values would also be economically advantageous for fishermen, as they would result in better catch quality, less time spent sorting, longer towing times, more catch, and cheaper fuel. With a shrimp retention of 96-77%, JFE-SSD operations off the coast of Cochin, India, have achieved a bycatch reduction of up to 43%.

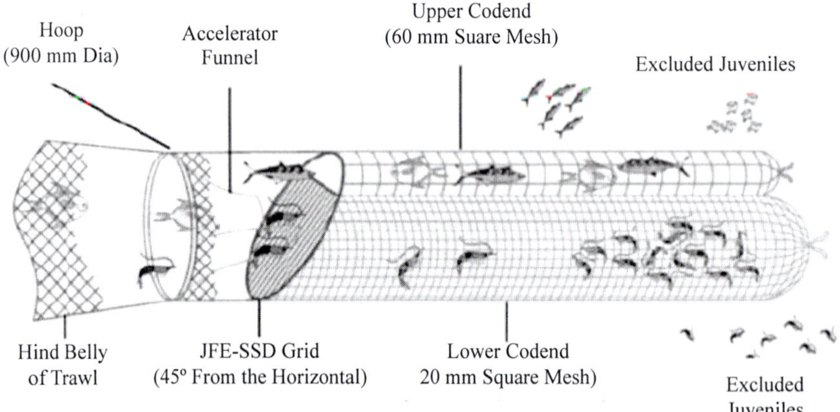

Fig. 7. Schematic diagram of JFE-SSD

10.1.7 CIFT-TED

The Central Institute of Fisheries Technology (Cochin, India) created an effective turtle excluder device known as CIFT-TED, with the goal of minimizing catch losses. However, trawler fishermen are hesitant to use the device because of this. The CIFT-TED is a straightforward, hard TED design with a top opening and a single grid. According to Dawson and Boopendranath (2001), catch losses during trial operations as a result of the installation of CIFT-TED were in the range of 0.52-0.97% for shrimp and 13.2-44-3.27% for catch components that were not shrimp. In cooperation with the corresponding state fisheries ministries and the MPEDA, CIFT-TED is now being made more widely known in maritime states. Despite the assurances provided by government agencies in India, the USA, Australia, and other countries that engage in shrimp trawling, fishermen maintain that the main obstacle impeding the widespread use of TEDs in India and other countries is the significant escape of shrimp from trawl nets.

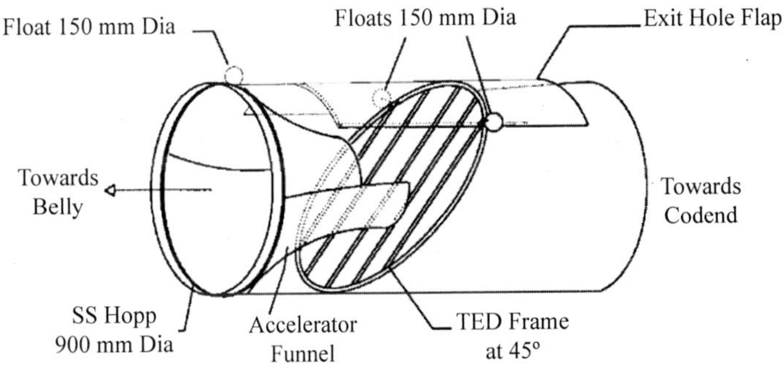

Fig. 8. Perspective diagram of CIFT-TED

10.2 Purse seine Fisheries

Purse seines are not selective, similar to the surrounding nets. Instead, schools were carefully chosen following an assessment of the species that may be caught as bycatch (operational selection). Bycatch species are regularly found in FAD-aided purse seining, and more than 40 species of fish and cetaceans have been documented in purse-seine landings (Romanov, 2002; Pravin *et al.*, 2008). Special escape panels called Medina panels, which are portions of thin mesh that prevent dolphins from being entangled in the gear, and back-down maneuvers have been placed to prevent the capture of dolphins in purse seines (Ben-Yami, 1994). Improved purse seine selectivity may also result from choosing a mesh size that is suitable for the target species, as well as from appropriate choices regarding the fishing region, depth, and season. Currently, there are no rules in place to reduce the number of protected species caught in purse seine fishing. One of the best ways to reduce accidental capture is to use fishery observers who monitor water before laying the nets.

10.3 Gillnet fisheries

In addition to non-targeted fish species, bycatch from drift gill nets includes marine mammals, sea turtles, and seabirds. Optimization of gill net mesh size and hanging coefficient according to the target species and size group and judicious deployment of gill nets in terms of fishing ground, fishing depth, and season to minimize gear interaction with non-targeted species are important bycatch mitigation measures for gill net fisheries. Recent improvements using acoustic pingers and carefully treated netting have sought to make gill nets observable by marine creatures with echolocation capabilities. Gill net fisheries employ acoustic pingers to prevent beaked whale bycatches (Carretta *et al.*, 2008). In some US, Northwest Atlantic, California, and European fisheries, using acoustic pingers is now required. It has been found that barium sulfate-treated acoustically reflecting polymaide netting reduces harbor porpoise bycatch in gill nets (Trippel *et al.*, 2003; Larsena *et al.*, 2007).

10.4 Ghost fishing

Since the material used to make gill nets is not biodegradable, lost nets continue to entangle gill fish and other marine life, resulting in unintended death. The practice known as "ghost fishing" is a drawback of contemporary gill nets; however, it is an easy, low-cost, low-energy fishing technique that works well for dispersed populations. Using time-release components or biodegradable natural fiber twines to join the netting to floats is one way to reduce ghost fishing caused by lost gill nets. When floats are detached owing to the dissolution of these linkages, the gill nets lose their fishing attitude and subsequently lose their capacity for ghost fish. Finding and recovering misplaced fishing equipment is another strategy for stopping ghost fishing.

10.5 Hook and line fisheries

The best ways to reduce gear interaction with other species and mitigate bycatch issues in hook and line fisheries are to optimize hook design and size, choose bait that is appropriate for the target species and size class, and choose fishing ground, depth, and time. By substituting circular hooks with traditional J-hooks in long-line operations, sea turtle mortality has decreased (Kerstetter and Graves, 2006). It has been observed that the use of weighted lines to set tuna longlines at depths beyond 100 m reduces the bycatch of fish species that are essential for recreational purposes and protected species, such as seabirds and sea turtles (Beverly *et al.*, 2009). Shark bycatch in long-line pelagic fishing has been suggested to be decreased by the magnetic field created by rare-earth magnets near the hook (Stoner and Kaimmer, 2008; WWF, 2009b).

10.6 Trap fisheries

Traps often require less energy to operate, have a high possibility of survival for rejected non-targeted species, and have good species specificity and size selectivity. Nonetheless, ghost fishing due to misplaced traps and a comparatively high loss rate during operation are drawbacks of trap fishing. Optimal trap design and mouth configuration in accordance with the target species, provision of escape windows for juveniles and non-target species on the design side, and appropriate selection of bait type, fishing area, fishing depth, fishing time, and fishing season on the operational side to minimize gear interaction with non-target species are some of the strategies to reduce bycatch in tap fishing.

10.6.1 Environment-friendly fishing gears

Certain fishing methods and equipment are more harmful than others, depending on how they affect ecosystems. The worst practices include the use of explosives and poisons, which are typically prohibited. Even though they have less of an effect on the environment, passive fishing gear, such as gillnets and traps, can lead to unreported fishing deaths due to ghost fishing from lost and abandoned fishing gear. Ragged gears, such as trawls, have the potential to seriously harm benthic plants and animals, which inhabit the bottom substratum and increase the productivity of the area, especially if they are severely rigged. When feasible, bottom trawls have been rigged to run a short distance above the sea floor, similar to semi-pelagic trawls, in an effort to decrease their influence on the substratum. Lines and large open pound nets (set nets) are two types of fishing gear that do not harm ecosystems.

10.6.1.1 Reducing bottom impact of towed gears

It is commonly recognized that bottom trawling has both direct and indirect effects on the marine environment and benthic populations (Hall, 1999; Kaiser and de Groot, 2000; Meenakumari *et al.*, 2009 and others). Benthic release panels, lighter gear design, semi-pelagic trawling, and minimization of the area of contact between the towed gear and the bottom are some of the improvements made to gear to meet the goal of decreased environmental effects (CEFAS, 2003; He, 2007).

10.6.1.2 Semi-pelagic trawl system

In addition to having an influence on benthic populations, demersal trawls are typically non-selective, landing on many non-target species and juveniles.

Resource-specific trawls targeting semi-pelagic resources have comparably modest impacts on benthic biota (He, 2007). After extensive field testing, the CIFT Semi-pelagic Trawl System (CIFT SPTS-I) was created as an alternative to shrimp trawling in the small-scale mechanized trawler sector (CIFT, 2007). It is capable of obtaining capture rates over 200 kg.h-1 on fairly productive grounds and selectively harvesting fast swimming demersal and semi-pelagic finfishes and cephalopods, which are typically outside the reach of conventional bottom trawls, currently utilized in commercial trawl fisheries in India.

10.6.1.3 Benthic release panels

Benthic release panels are large drop-out windows made of square mesh placed ahead of the codend to allow undesirable benthic animals to be released (Fonteyne and Polet, 2002; He, 2007).

10.6.1.4 Remote controlled otter boards (smart trawling)

The trawl is operated at a predetermined height above the sea bottom without causing any bottom impact owing to variable thrust vector devices (VTVD), which are based on the Magnus effect (towed rotating cylinders generate lateral forces perpendicular to the axis of the cylinder and towing direction) (Shenker, 2005). Another option in this area is the acoustic control of otter boards (CEFAS, 2003; He, 2007).

10.6.1.5 Ground gear modifications

It has been shown that using lighter ground gear during trawling might lessen the bottom impact without appreciably impacting the capture rates (He, 2007). In addition to the benefits of fuel savings from a reduction in trawl drag, a technique to reduce the bottom effect of a trawl gear has been proposed: rollers, wheels, and bobbins with axes perpendicular to the towing direction (He, 2007).

10.6.1.6 Otter boards bridles and sweeps

High proportion of aspects in comparison to classic otter boards, vertically cambered otter boards usually have a lower angle of attack and a thinner footprint (He, 2007). When otter boards with a high aspect ratio are placed on a comparable board area, the area of the seafloor that they influence is usually 40% larger than that of otter boards with a low aspect ratio. When the herding effect is not crucial to the capture process, using lighter and shorter bridles and sweeps might lessen the impact on the seafloor.

11. Energy conservation in fishing

One of the most energy-intensive methods to produce food is fishing. Gillnets, entangled nets, lines, and traps are examples of passive fishing techniques and equipment that require less energy than active dragging fishing techniques (Gulbrandson, 1986; Boopendranath, 2009). Trawling is one of the fishing gears that uses the most energy per unit quantity of catch and provides more opportunities for energy-saving techniques. Because purse-seining lands high quantities of fish per operation, it uses significantly less energy per unit quantity of catch landed. This may entail actions such as choosing and promoting low-energy fishing methods when practical, implementing energy-saving techniques and equipment in energy-intensive fishing systems when necessary, utilizing cutting-edge technologies such as remote sensing, acoustic fish detection, and global positioning systems to locate fishing grounds more accurately and save time spent searching for fish, and implementing measures to develop and enhance coastal fishing grounds so that fishing can occur in nearby waters rather than far-off ones.

Fishing, which is motorized and mechanized, depends on fossil fuels that are finite and non-renewable. Increased atmospheric carbon dioxide levels from fossil fuels contribute to the greenhouse effect and other pollutants that are harmful to the environment and public health. Oceanic and climatic patterns are irreversibly altered by the greenhouse effects. Moreover, the economic sustainability of fishing as a food production method may be significantly affected by the skyrocketing cost of oil. Taking these consequences into account, several countries have launched extensive energy conservation programs. An estimated 134 million tonnes of CO_2 are released into the atmosphere each year by the world's capture fisheries, which use around 50 billion liters of fuel yearly (or 1.2% of the world's fuel consumption) at an average rate of 1.7 tons of CO_2 per tonne of live-weight landed product (Tyedmers *et al.*, 2005). Gulbrandson (1986), Wilson (1999), Boopendranath (2009), and others have evaluated a variety of methods for energy saving in fish harvesting, including the following:

- Low energy fishing techniques
- Low drag trawls
- Pair trawling
- Economic vessel speed
- Hull design and displacement optimisation
- Anti-fouling measures
- Choice of engines

- Right sizing of engines
- Emission standards
- Preventive maintenance of engines
- Reduction gear, propeller size and propeller nozzle
- Sail-assisted propulsion
- Use of advanced technology such as Echosounder, Sonar, GPS, PFZ information, GIS, for fish finding and navigation
- Fleet management

Conclusion

The introduction of powerful and highly efficient fish harvesting systems and fish detection methods has increased fishing efficiency several times in recent years. Recent developments in fishing technology have focused on reducing the bycatch of non-target species, protected species, and juveniles, minimizing the environmental impact of fishing gear and their operation, and minimizing the energy use per unit volume of fish landed during fishing operations. A wide range of proven technologies are readily available for adoption under the responsible fishing regime in the areas of bycatch reduction, mitigation of negative environmental impacts, and conservation of energy in fishing. The degree of adoption of responsible fishing technologies strongly depends on the robustness of the fisheries management system. Technologies that have inherent advantages in terms of net profit, such as some fuel conservation measures (e.g., low-drag trawl systems), have been more easily adopted than others. Bycatch reduction technologies and mitigation measures for the environmental impacts of fishing have been mandated and effectively implemented in several scientifically managed fisheries worldwide. However, its adoption in less effectively managed fisheries with an open access regime may require the active involvement of stakeholders in the process, supported by a system of incentives, disincentives, and training. A rights-based regulated access system based on a strong inclusive participatory system of management seems necessary for facilitating the large-scale adoption of responsible fishing technologies.

References

Alverson, D.L., Freeberg, M. H., Murawski, S. A. and Pope. J.G., (1994). A Global assessment of fisheries bycatch and discards. FAO Fish. Tech. Pap. No 339. Rome, FAO, 233 p.

Ben-Yami, M. (1994) FAO Purse Seining Manual, Fishing News Books Ltd., UK: 406 p.

Beverly, S., Curran, D., Musyl, M., and Molony, B. (2009). Effects of eliminating shallow hooks from tuna longline sets on target and non-target species in the Hawaii-based pelagic tuna fishery Fisheries Research 96 (2-3): 281-288

Boopendranath, M.R. (2008). Climate change impacts and fishing practices, Paper presented in Workshop on Impact of Climate Change in Fisheries, 15 December 2008, ICAR, New Delhi.

Boopendranath, M.R. (2009). Responsible fishing operations. In: Handbook of Fishing Technology (Meenakumari, B., Boopendranath, M.R., Pravin, P., Thomas, S.N. and Edwin, L. (Eds), Central Institute of Fisheries Technology, Cochin: 259-295.

Boopendranath, M.R. and Pravin, P. (2009). Technologies for responsible fishing - - Bycatch Reduction Devices and Turtle Excluder Devices. Paper presented in the International Symposium on Marine Ecosystems-Challenges and Strategies (MECOS 2009), 9-12 February 2009, Marine Biological Association of India, Cochin.

Carretta, J.V., J. Barlow, and L. Enriquez, (2008). Acoustic pingers eliminate beaked whale bycatch in a gill net fishery. Marine Mammal Science, 24(4), ps 956- 961.

CEFAS, (2003). A Study on The Consequences of Technological Innovations in Capture Fishing Industry and the Likely Effects upon Environmental Impacts. Submitted to the Royal commission on Environmental Pollution, London, UK Centre for Environment, Fisheries and Aquaculture, Lowestoft, UK, 181 pp.

Cheung, W. W. L., Sarmiento, J. L., Dunne, J., Frölicher, T. L., Lam, V. W. Y., Palomares, M. L. D., et al. (2013). Shrinking of fishes exacerbates impacts of global ocean changes on marine ecosystems. Nat. Climate Change 3 (3), 254–258. doi: 10.1038/nclimate1691.

Chokesanguan, B., Ananpongsuk, S., Siriraksophon, S., Podapol, L. 2000. Study on Juvenile and Trash Excluder Devices (JTEDs) in Thailand. SEAFDEC Training Department, Samut Prakan, Thailand, TD/RES/47, 8 pp.

CIFT (2007). Responsible Fishing - Contributions of CIFT, CIFT Golden Jubilee Series, Central Institute of Fisheries Technology, Cochin: 46 p

Courtney, A. J., Haddy, J. A., Campbell, M. J., Roy, D. P., Tonks, M. L., et al. (2007). Bycatch weight, composition and preliminary estimates of the impact of bycatch reduction devices in Queensland's trawl fishery. FRDC 2000/170 Final Report.

Dawson, P. and Boopendranath, M. R. (2002). Application of CIFTTED for turtle conservation. In: Proceedings of the Workshop on the operation of Turtle Excluder Device (TED), 24-25 January 2002, Kakinada, Department of Fisheries, Andhra Pradesh, p. 12-14.

FAO (2011). International guidelines on bycatch management and reduction of discards. Food and Agriculture Organisation of the United Nations. FAO, Rome, p 73Return to ref 2011 in article.

FAO 1995. Code of Conduct for Responsible Fisheries, FAO, Rome, 41 p.

FAO 1996. Fishing Operations, FAO Technical Guidelines for Responsible Fisheries 1, 26 pp.

Fonteyne, R. and Polet, H. (2002). Reducing the benthos bycatch in flatfish beam trawling by means of technical modifications. Fish. Res. 55: 219-230

Funge-Smith S, Lindebo E, Staples D. 2005. Asian fisheries today: The production and use of low value/trash fish from marine fisheries in the Asia-Pacific region. FAO Regional Office for Asia and the Pacific. RAP Publication 2005/16, 38 p.

Gilman, E., Brothers, N. and Kobayashi, D. (2003). Performance assessment of underwater setting chutes, side-setting, and blue-dyed bait to minimize seabird mortality in Hawaii pelagic longline tuna and swordfish fisheries. Final Report. U.S. Western Pacific Regional Fishery Management Council, Honolulu

Gilman E, Pérez Roda MA, Huntington T, Kennelly SJ, Suuronen P, Chaloupka M, Medley PAH (2020). Benchmarking global fisheries discards. Nat Sci Rep 10:14017. https://doi.org/10.1038/s41598-020-71021-xReturn to ref 2020 in article.

Glass CW, Walsh SJ, van Marlen B, Amaratunga T (2007). Fishing technology in the 21st century: integrating fishing and ecosystem conservation. ICES J Mar Sci 64(8):1499–1616.

Gulbrandson, O. (1986). Reducing Fuel Cost of Small Fishing Boats, BOBP/WP/27, Bay of Bengal Programme, Madras:15 pp.

Hall, S.J. (1999). The effect of fishing on marine ecosytems and communities, Blackwell, Oxford, UK, 244 pp.

He, P. (2007). Reducing seabed contact of bottom trawls. In: Proc. Satellite workshop on fishing Impacts – Evaluation, solution and Policy, October 2001, Tokyo, Japan, 27-35.

Heppell, S. S., L. B. Crowder, D. T. Crouse, S. P. Epperly, and N. B. Frazer. (2003). Population models for Atlantic loggerheads: past, present, and future. Pages 255–273 in A. B. Bolton and B. L. Witherington, editors. Loggerhead sea turtles. Smithsonian Books, Washington, D.C.

Hilborn, R., Amoroso, R., Collie, J., Hiddink, J. G., Kaiser, M. J., Mazor, T., et al. (2023). Evaluating the sustainability and environmental impacts of trawling compared to other food production systems. ICES J. Mar. Sci. 80 (6), 1567–1579. doi: 10.1093/ icesjms/ fsad115.

ICES (2020). ICES workshop on innovative fishing gear (WKING). ICES Sci. Rep. 2 (96), 130 pp. doi: 10.17895/ices.pub.7528.

Isaksen, B., Valdemarsen, J.W., Larson, R.B., Karlsen, L., (1992). Reduction of fish bycatch in shrimp trawls using a rigid separator grid in the aft belly. Fish. Res. 13: 335-352.

Johnsen JP (2005). The evolution of the ''harvest machinery'': why capture capacity has continued to expand in Norwegian fisheries. Mar Policy 29(6):481–493.

Kaiser, M.J. and de Groot, S.J. (2000). Effect of fishing on non-target species and habitat, J. Anim. Ecol. 65: 348-358

Kasperski, S., and Holland, D. S. (2013). Income diversification and risk for fishermen. Proc. Natl. Acad. Sci. 110 (6), 2076–2081. doi: 10.1073/pnas.1212278110.

Kennelly, S. J., and Broadhurst, M. K. (2021). A review of bycatch reduction in demersal fish trawls. Rev. Fish Biol. Fisheries. 31 (2), 289–318. doi: 10.1007/s11160-021- 09644-0.

Kerstetter, D. W., and J. E. Graves. (2006). Effects of circle versus J-style hooks on target and non-target species in a pelagic longline fishery. Fisheries Research 80:239-250.

Larsena, F., Eigaarda, O.R. and Tougaardb, J. 2007. Reduction of harbour porpoise (Phocoena phocoena) bycatch by iron-oxide gillnets Fisheries Research 85(3): 270-278.

Lewison, R. L., Crowder, L. B., Wallace, B. P., and Safina, C. (2014). Global patterns of marine mammal, seabird, and sea turtle bycatch reveal taxa-specific and cumulative megafauna hotspots. Proc. Natl. Acad. Sci. 112 (20), 6210–6215. doi: 10.1073/ pnas.1318960111.

Lucchetti, A., Melli, V., & Brčić, J. (Eds.). (2023). Innovations in fishing technology aimed at achieving sustainable fishing. Frontiers Media SA.

Meenakumari, B., Bhagirathan, U. and Pravin, P. (2009). Impact of bottom trawling on benthic communities: a review, Fish. Technol. 45(1):1-22.

Nguyen BT. (2017). Study on trawl fishery socio-economics and supply chains in Kien Giang, Viet Nam Pp. 175– 237. In: Siar S. V., P. Suuronen, and R. Gregory, Eds. Socio-economics of trawl fisheries in Southeast Asia and Papua New Guinea. FAO Fisheries and Aquaculture Proceedings No. 50. FAO Bangkok.

Nuhu MB, Yaro I (2005). Selection of efficient hanging ratios of gillnet on fish catch in Lake Kainji, as a means of alleviating poverty among artisanal in Nigeria. In Araoye PA (Ed). Proceedings of the 19th Annual Conference of the Fisheries Society of Nigeria (FISON). P.64- 72.

Polet, H., Coenjaerts, J. and Verschoore, R. (2004). Evaluation of the sieve net as a selectivity-improving device in the Belgian brown shrimp (Crangon crangon) fishery. Fish. Res. 69: 35-48.

Pravin, P., Meenakumari, B. and Boopendranath, M.R.(2008). Harvest Technologies for Tuna and Tuna like fishes in Indian seas and bycatch issues. In: Harvest and Post-harvest Technology for Tuna (Joseph J., Boopendranath M.R., Sankar T.V., Jeeva J.C. and Kumar R., (Eds)), Society of Fisheries Technologists (India), Cochin, 79-103 pp.

Revill, A. and Holst, R. (2004). The selective properties of sieve net, Fish. Res. 66: 171-183

Robins, J.B., Campbell, M.J. and McGilvrey, J.G. (1999). Reducing prawn trawl bycatch in Australia: An overview and an example from Queensland. Marine Fisheries Review 61(3): 46-55.

Romanov, E.V. (2002). Bycatch in the tuna purse-seine fisheries of the western Indian Ocean, Fish. Bull. 100(1): 90–105.

Squires, D., and Vestergaard, N. (2013). Technical change in fisheries. Mar. Policy. 42, 286–292. doi: 10.1016/j.marpol.2013.03.019.

Stoner, A. W., and S.M. Kaimmer., (2008). Reducing elasmobranch bycatch: laboratory investigation of rare earth metal and magnetic deterrents with spiny dogfish and Pacific halibut.. Fish. Res. 92:162-168.

Tidd A, Brouwer S, Pilling G (2017). Shooting fish in a barrel? Assessing fisher-driven changes in catchability within tropical tuna purse seine fleets. Fish Fish 18(5):808–820.

Trippel, E.A, Holy, N.L. Palka, D.L. Shepherd, T.D. Melvin, G.D. and Terhune, J.M. 2003. Acoustic reflective net mesh reduces harbour porpoise bycatch. Marine Mammal Science. 19: 240-243.

Valdemarsen, J. W. (2001). Technological trends in capture fisheries. Ocean & Coastal Management, 44(9-10), 635–651. doi:10.1016/s0964-5691(01)00073-4

Wang SG, Wang RZ (2005) Recent developments of refrigeration technology in fishing vessels. Renew Energy 30(4):589–600.

Watling, L., and Norse, E. A. (1998). Disturbance of the seabed by mobile fishing gear: a comparison to forest clear cutting. Conserv. Biol. 12 (6), 1180–1197. doi: 10.1046/j.1523-1739.1998.0120061180.x.

Wilson, J.D.K. (1999). Fuel and Financial Savings for Operators of Small Fishing Vessels, FAO Fish. Tech. Paper 383, FAO, Rome.

WWF. (2009a). Modifying shrimp trawls to prevent bycatch of non-target species in the Indian Ocean, Accessed 20 May 2009, www.smartgear.org/ smartgear_winners/smartgear_winner_2005/smartgear_winner_2005.

WWF 2009b. Deterring sharks with magnets, Accessed 20 May 2009, http://smartgear.org/smartgear_winners/smartgear_winner_2006/smartgear_ winner_2006grand/index.cfm.

11

Fish Hatchery
Good Business for Doubling Income

Prabir Sahoo[1*], Sagar Samanta[2] and Pijush Payra[3]

[1]*Department of Coastal Aquaculture (M.Sc. student), Centre of Advanced Study in Marine Sciences, Annamalai University, Cuddalore-608502, Tamil Nadu, India*

[2]*Department of Fisheries Science (M.Sc. Student), Medinipur City College West Medinipur-721129, West Bengal, India*

[3]*Department of Industrial Fish and Fisheries (HOD), ACM and Fisheries Science (PG), Ramnagar College, Depal-721453, West Bengal, India*

Abstract

For business owners seeking to diversify their revenue sources, aquaculture, especially fish hatchery operations, has become a lucrative venture. Fish farming has emerged as a crucial link in the food supply chain owing to the rising demand for seafood worldwide and the loss of wild fish supplies. To double the income, the possibility of creating a fish hatchery was investigated in this study. This report demonstrates the feasibility of fish hatchery enterprises in producing significant returns on investment through a thorough examination of market trends, operational concerns, and financial projections. To optimize profitability in this industry, the results highlight the need to utilize technological breakthroughs, embrace environmentally friendly behaviors, and implement successful marketing plans.

Keywords: *fish hatcheries, aquaculture, operational efficiency, sustainable practices, technological advancements, marketing strategies.*

1. Introduction

Starting a fish hatchery has the potential to be a profitable endeavor with substantial income growth. Fish hatcheries are essential to the aquaculture sector because they breed and raise fish for a variety of uses, including commercial operations, recreational fishing, and stocking. This business has numerous advantages such as the opportunity for sustainable production methods, consistent demand for fish products, and very low initial investment expenses. Furthermore, fish hatcheries offer a practical way to address the growing global demand for seafood while also promoting economic development and job creation in nearby communities. The context for discussing the benefits and potential of establishing a fish hatchery as a way to double revenue was established by this introduction. In an established fish hatchery, fish eggs are produced and reared under carefully monitored circumstances in preparation for their eventual release into the wild or transfer to aquaculture farms. The potential of fish hatcheries to double income is examined in this introduction, which also emphasizes important aspects such as market demand, sustainability, technological improvements, and financial feasibility.

2. History of hatchery evaluation

2.1 Origins of the 19th century (1800s)

- Fish populations dropped as a result of overfishing and habitat damage, which led to the emergence of fish hatcheries in the 19th century.
- The first federally funded fish hatchery in the United States was founded in 1833 at the Baird Station on the McCloud River in California. New York State was the founder.

2.2 Development and Growth (Late the 1800s–Early the 1900s).

- Fish hatcheries spread throughout the United States and other nations in the late 19th and early 20th centuries, propelled by developments in aquaculture technology and rising awareness of the need for fish conservation.

The United States Fish Commission was founded by the US Congress in 1871 and was instrumental in the growth and development of fish hatcheries throughout the country. Fish hatcheries grow much faster at this time because of the emergence of artificial propagation methods, including controlled breeding and artificial insemination.

2.3 Industrialization and Modernization (Mid-20th Century)

The mid-20th century saw significant advancements in fish hatchery technology, including the development of large-scale hatcheries capable of producing millions of fish fry annually.

Many governments around the world have begun investing heavily in fish hatcheries as part of broader efforts to enhance food security and support commercial fishing industries.

Research into selective breeding and genetic improvement techniques has become increasingly important, leading to the production of fish with desirable traits, such as fast growth rates and disease resistance.

2.4 Environmental Concerns and Sustainable Practices (Late 20th Century–present)

In recent decades, fish hatcheries have faced criticism for their potential negative environmental impacts, including genetic introgression, disease transmission, and disruption of natural ecosystems.

As a response, there has been a growing emphasis on sustainable aquaculture practices, including the implementation of strict regulations, habitat restoration efforts, and the adoption of environmentally friendly technologies.

Many modern fish hatcheries now prioritize conservation and environmental stewardship, aiming to balance the needs of commercial fishing with the preservation of natural ecosystems.

2.5 Trends and Opportunities in the 21st Century

Fish hatcheries play a critical role in aquaculture production, fisheries management, and conservation activities worldwide. Given the continually rising demand for seafood worldwide, there is considerable potential for fish hatcheries to grow and diversify their operations, including the production of high-value species for niche markets and the development of cutting-edge aquaculture technologies. Furthermore, fish hatcheries can act as catalysts for economic development in rural areas by creating jobs and boosting local economies by selling fish and associated goods.

3. What is hatchery?

A fish hatchery is a facility where fish are reared in controlled environments after hatching. These facilities are usually utilized for a variety of purposes, such as fishing rivers, lakes, and oceans for commercial, recreational, or conservation purposes.

3.1 This was a thorough explanation

i) Egg Collection: Mature female fish are the first source of eggs for hatcheries. Depending on the species, this may entail physically harvesting eggs or using specialized equipment.

ii) Fertilization: After gathering, male fish sperm fertilizes the eggs. In some circumstances, this can happen naturally, whereas in others, hatchery workers can assist.

iii) Incubation: After fertilization, the eggs were placed in trays or tanks that had been properly made with regulated water. These circumstances are similar to the natural conditions required for egg development.

iv) Hatching: The eggs eventually hatch into fry or freshly hatched fish. Depending on the species and surroundings, this process can take several days or weeks.

v) Nursery Rearing: The fry are placed in ponds or tanks after hatching, where they receive nourishment and attention until they reach a specific size and developmental stage.

vi) Growth: The fish continue to grow until they are ready for harvest or stocking after reaching a particular size, at which point they may be moved to larger tanks or ponds, or even released into natural bodies of water.

vii) Monitoring and Management: To guarantee the health and welfare of fish during the entire process, hatchery staff feed the fish, check the water quality, and administer it.

4. Why set up a Fish Hatchery

4.1 Examining the Market

Target Market Identification: Determine which particular market groups you will serve, such as hobbyists of aquariums, recreational pond fishers, or commercial fish farms. This is known as target-market identification.

Competitor Analysis: Analyze the offerings, costs, and market share of fish hatcheries that are currently operating in the area.

Trends and Growth: Market trends, such as the rising demand for sustainably produced food, should be investigated, as they have the potential to propel the expansion of fish farming enterprises.

Regulatory Environment: Environmental and Health Regulations: Recognize the laws and rules controlling fish hatcheries, such as permits and environmental requirements.

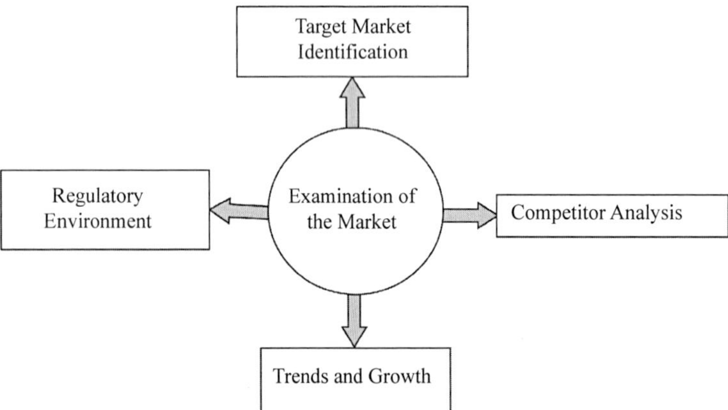

4.2 Demand

Increasing Consumption: Draw attention to the rising need for fish and seafood products as a result of dietary changes, health consciousness, and population expansion.

Supply Shortages: Overfishing and environmental reasons may lead to shortages in fish supplies. To address this issue, opportunities should be created for fish hatcheries to meet the demand.

Diversification of Species: Choose popular fish species, such as tilapia, salmon, or trout, and adjust hatchery practices to suit their needs.

Local and Sustainable Sourcing: Appeal to consumers' preferences for ethical and ecologically friendly products by highlighting the benefits of seafood obtained locally and produced responsibly.

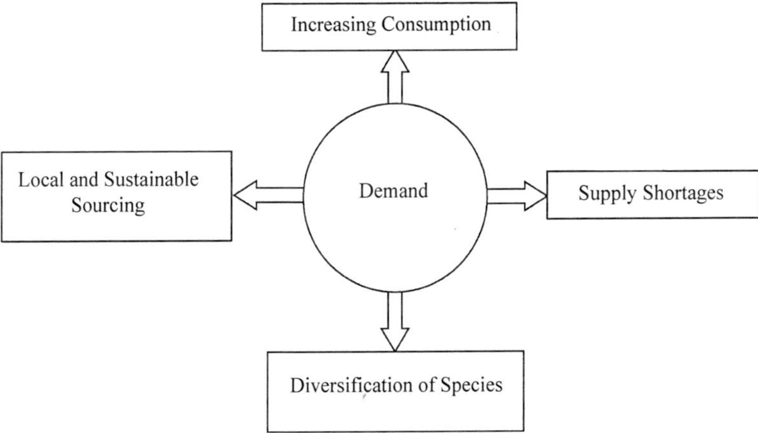

4.3 Cost Savings

Producing fingerlings in a hatchery is more cost-effective than purchasing them from external sources. This cost-saving measure could contribute to higher profits for fish farmers.

4.4 Increased Production

Fish hatcheries produce high-quality fingerlings (young fish), which can be raised on fish farms to increase production. With more fish available for sale, farmers can earn higher revenue.

4.5 Control over Quality

Fish hatcheries allow farmers to have better control over the quality of the fingerlings they raise. By ensuring the health and genetics of fish, farmers can produce more robust and desirable fish for the market, commanding better prices.

4.6 Risk Management

Hatcheries provide a degree of risk management by reducing reliance on wild-caught fingerlings, which can be subject to environmental factors and fluctuations in availability. By producing their fingerlings, farmers can mitigate these risks and maintain consistent production levels.

5. Types of hatchery

5.1 Earthen Pot Hatchery

i) An overhead tank in the shape of an earthen clay pot was placed on a brick platform, approximately two feet high, in accordance with the hatchery system concept.

ii) However, the two containers were not of the same size.

iii) The first container served as an incubator and the second container served as a spawnary.

iv) A rubber pipe and a valve were attached to the bottom of the clay pot.

v) A metal tube is located at the opposite end of the rubber pipe and is placed inside container I.

vi) A circular mesh sheet was positioned and secured at the top of the container by using a bamboo rod.

vii) A circular spout is installed close to the surface at the other end of container I.

viii) The opening of this spout from container I leads to container II, which is smaller than the first one.

ix) An overflowing pipe was installed near the surface of container II's other end, and a net was positioned on top of it.

x) 10.A small net was placed on the outside of the pipe.

5.1.1 Operational Procedure

i) First, freshwater was added to the clay pot in a tube well. The valve was controlled to preserve the water pressure entering container I.

ii) Fertilized eggs were added to the containers.

iii) The extra water exits the system by the spout, fills container I1, and then uses the overflowing pipe to go outdoors.

iv) Thus, the running water condition was maintained.

5.2 Glass Jar Hatchery

5.2.1 Location

i) A glass jar hatchery is typically set up next to a water source, which can be a tube well, river, or perennial pond.

ii) It must be stored in a room with sufficient light, ventilation, and a suitable drainage system.

5.2.2 Construction

5.2.2.1 Overhead Tank

i) The source's available water was pumped into an overhead tank situated above a brick wall.

ii) If water is extracted from a river, a desilting tank must be included. The pump system was finely meshed by wire-netting.

iii) This aids in preventing the introduction of organisms frequently found in rivers and pond water.

iv) A deep tube well is useful because it can supply pure water.

v) Installing the tube well in an area free of arsenic is recommended.

5.2.2.2 Breeding Tank

i) Hatcheries are self-contained facilities that often include breeding tanks equipped with overhead shower systems.

ii) Fish were meant to spawn in this tank. Every tank had an exit to allow extra water to be drained.

iii) A breeding tank has the following measurements: 1.8 m x 0.9 m x 0.9 m.

iv) Typically, a farm has two to three breeding tanks with the capacity to fix two to four breeding hapas simultaneously.

v) However, the spawn production target determines the number of breeding tanks that a hatchery needs.

vi) When a hatchery lacks breeding tanks, the process is carried out in hapas set in ponds, and the fertilized eggs are released into glass jars for incubation and hatching.

5.2.2.3 Incubation and Hatching Jars

A variety of jars, including glass, synthetic, and Zong jars with temperature-controlled water sources, were used for incubation and hatching.

5.3 Glass

1. The glass jars have a conical shape.
2. Their tops were open, and as they moved closer to the bottom, they tapered.
3. Jar's measurements.
4. A hardwood table has circular holes drilled into it, through which glass jars are arranged vertically.

5. The table was 4.05 m × 0.6 m and had a 0.9 m height.
6. Meticulous strategy for arranging jars on a table in order.
7. There is a 22.9 cm space between each of the two rows of jars.
8. The clamps assist and maintain the clamps vertically.
9. A beak-shaped ring made of galvanized iron was attached to the top of the jar.
10. It acts as a vent for water leaking from the jar.
11. Rubber tubing with corresponding taps was used to connect the jars at the bottom.
12. A total of 50,000 carp eggs can be stored in each 6.35-liter glass jar at a time.
13. Water from each jar overflows into a long open galvanized tube, which is installed such that the water runs down to the spawnary.

5.4 Zug Jars

1. The Zug jars, named Zug, Switzerland, the place where they were originally employed, are large, inverted bottles with open tops and narrow bottoms that measure between 60 and 70 cm in height and 15-20 cm in diameter.
2. They are arranged vertically in a hatchery in succession, with the narrow neck pointing downward and connected to the water pipes using appropriate taps.
3. The outflow was located at the top, and water was supplied from below.
4. Zug jars hold roughly 1–5 liters of eggs and have a 6–8 liter capacity.

5.5 MacDonald or chase jars

Zug Jars and MacDonald Jars were comparable. It has a circular water supply bottom and cylindrical shape, measuring 40–50 cm in height. Water flow in the jar causes the eggs to rotate gently.

5.6 Plastic Jars

1. Low-density polythene was used to make these jars.
2. The upper section had a capacity of 40 liters in capacity and diameters of 440 mm and 625 mm, respectively.
3. Each jar has a separate control valve. After hatching, the eggshells were effectively separated using an internal egg container.
4. When the intake pipe reached deep into the bottle, a hatching jar equipped with a water supply was utilized.

5.2.2.4 Spawnery

1. The spawnery is a synthetic material hapa that is installed on a frame and kept inside a tank or a concrete cistern.
2. To prevent the hatchlings from being carried away with the spilling water from the cistern, the hapa is somewhat taller than the concrete cistern.
3. Water was sprayed over the hatchlings growing in the hapa using an overhead shower.
4. The hapa is 1.65 meters by 0.8 meters by 1.0 meters, while the spawnery's cement cistern is 1.8 meters by 0.9 meters by 0.9 meters.
5. It has a capacity to spawn approximately 10 lakh at a time.

5.2.2.4.1 Working Method

1. Water is pumped into the tank from nearby ponds, rivers, or groundwater.
2. Water flowed into the spawnery and incubation jars from the above tank.
3. Subsequently, approximately 50,000 water-hardened eggs were gradually added to each incubation jar. Throughout the incubation process, the water pressure was controlled to ensure that the eggs swirled gently without breaking or spilling over.
4. It requires 600–800 ml of flow per minute to maintain the carp eggs moving inside the jar.
5. Through an opening at the bottom, water can enter the incubation jars from above or from the bottom.
6. Following hatching, the eggs exhibit a tendency for vertical migration, which causes the hatchlings to rise to the surface and flow through the outlet with the overflowing water into the open conduit that leads to spawning. After hatching was completed, the water flow rate in the jars was slightly increased to help the hatchlings escape as quickly as possible.

   The jars were then empty, and aberrant and dead hatchlings, unfertilized eggs, dead eggs, and eggshells were removed. This was done by unplugging the glass or incubation jar from the tap.

   When the fry in the spawnery is ready for stocking, water is sprayed constantly for approximately three days.

5.7 Merits and Demerits of Glass Jar Hatchery

5.7.1 Merits

1. Its operation is possible within a small space.
2. Developing embryos may be seen with the unaided eye in transparent incubation jars, allowing for the necessary corrections to be made.

5.7.2 Demerits

1. Glass incubation jars are susceptible to breakage and cannot be moved to other locations.
2. No temperature control system is available.
3. There is no additional airflow.
4. The hatching success rate was lower than that in Chinese hatcheries.
5. There is a chance that eggs from glass jars will be damaged during transfer.

6. Chinese Hatchery (eco-hatchery).

Water continuously flows in a circular pattern inside circular hatching chambers created by Chinese researchers. This system is commonly referred to as the Chinese hatchery system.

When raising carps for commercial carp seed production is needed, this approach has proven to be an extremely effective way to replicate some characteristics of the riverine ecosystem in a short space. Incubation of the hatching tank, spawn-rearing tank, spawning pool, and overhead tank make up the Chinese hatchery system.

6.1 Construction

6.1.1 Overhead Tank and Spawning Pool or Breeding Tank.

These procedures have been described previously.

6.1.2 Incubation of Hatching Tank

The hatching tank was made of bricks and cement and had a circular shape. It consisted of two concentric circular tanks. The interior diameters of the two chambers were 1.6 and 5.0 meters. The inner circular tank had an output pipe installed in the middle that maintained a water depth of 0.92 1.0 m.

Water enters the inner space through holes in the body of the inner circular tank. This results in a double-doughnut-like form. The double doughnut has two walls: one on the outside and the other on the inside, which encircles the exit.

A fine nylon screen was stretched and placed on an iron frame to divide the inner wall or chamber from the exterior before the eggs were introduced. The chamber was sealed using an extremely tight rubber belt.

When the taps are opened, water circulates (clockwise) through a series of diagonally arranged pipes on the outer compartment floor facing one way.

6.1.3 Spawning Tank

After the eggs hatched, the hatching tank served as the spawning tank, ensuring that water flow was maintained at its ideal level. Here, the hatchlings spent three days raising.

6.2 Working Principle

Male and female brooders were released into the spawning pool after pituitary injection. Before the eggs were added, the hatchery was prepared and the output pipe was adjusted to maintain the water level.

The spawning pool's eggs can be mechanically moved or drained through an underground pipe system straight into the hatchery's outer chamber.

The circular circulation condition created by the water arrangement of circulation in the floor-mounted pipes allows the eggs in the outer chamber to be kept there. Continuous churning of the eggs and high oxygen content in the water were made possible by the incubation tank.

The selected water flow rate contributed to maintaining hatching success. Because of elevated oxygen levels during hatching, a high flow rate requires an increase in temperature. The embryo begins to hatch approximately 11–12 h after conception and is completed in 4 h.

The water flow rate was lowered once every egg had hatched. Egg shells, unfertilized eggs, dead hatchlings, etc., are routinely removed from the incubation chamber to prevent bacterial breakdown.

Either siphoning or dipping a jute rope into the outer chamber while it is weighed accomplishes this. The jute rope, which is connected to the circulating eggshells, is removed and replaced with eggshells until almost all of the eggshells have been eliminated.

In hatcheries, some unfertilized eggs are left behind, and they serve as the major target for fungal infections caused by Saprolegnia. Applying malachite green for 20 to 25 min after turning off the water flow at a concentration of 0.02 g/liter of water is a typical technique for managing fungal infections in the incubator. The chemicals were removed from the incubator after the flow was restarted.

6.3 The Merits and Demerits of Chinese Hatcheries

6.3.1 Merits

1. Chinese hatcheries guarantee a high hatching success rate (98%) due to the possibility of good and clean water replacement.
2. During hatching, the water flow and temperature were maintained.
3. There were direct connections between the Chinese hatchery and nursery pond, as well as the breeding pool. Consequently, it guarantees less harm to spawning and fertilized eggs during transfer.
4. It lowers labor costs.
5. A large number of eggs can be held.

6.3.2 Demerits

1. It is impossible to completely remove shells, unhatched eggs, dead spawns, and larvae because of the lack of suitable spawnery.
2. The setup has a fairly high initial cost.
3. The amount, direction, and fitting of pipes must be performed with proper technical applications.

7. The following points are important for doubling income

i) High Demand: There steady market for fish goods owing to the rising worldwide demand for food. Maintaining fish hatcheries guarantees consistent supply to satisfy this need.

ii) Profitability: Compared to other forms of animal husbandry, fish farming can be profitable with comparatively low overhead costs. Maximizing earnings through effective resource and production management.

iii) Varied Income Streams: Fish hatcheries can make money in a number of ways, such as by selling fingerlings to different fish farms, providing fish to nearby markets or dining establishments, or even providing ecotourism possibilities, such as guided hatchery tours.

iv) Scalability: A hatchery can grow in a manner that is controlled by starting small and progressively expanding. Acquire resources and experience and then expand production to boost revenue.

v) Environmental Benefits: Sustainable aquaculture methods and less strain on wild fish populations are two ways in which ethical fish farming can benefit the environment. Customers who care about the environment may consider this to be a selling point.

vi) Technology Integration: Processes can be optimized, efficiency can be increased, and labor costs can be decreased by integrating technology into hatchery operations. Data analytics, water quality monitoring, and automated feeding systems can help maximize revenue.

vii) Government Support: To promote fish farming as a sustainable food supply, several governments provide grants, subsidies, or other forms of financial assistance. Utilizing these initiatives can increase profitability and reduce setup expenses.

viii) Diversification: Including a fish hatchery in an existing agricultural company can help distribute risk, increase overall profitability, and diversify revenue streams.

ix) Long-Term Investment: Because fish have shorter production cycles than other animals, fish farming is generally a long-term investment that can yield steady income for many years.

x) Adaptability: By modifying the type of fish they produce or concentrate on niche markets, such as organic or specialty fish products, hatcheries may adjust to shifting market conditions and consumer preferences.

8. Market strategies for doubling the income with fish hatchery

8.1 Optimizing Production

Increase efficiency: boost productivity by examining present processes and identifying opportunities for development. This could entail improved feed usage, disease control techniques, or water quality management.

Make an automation investment: Automate time-consuming chores such as watering and feeding to free up time for more important work and possibly reduce labor expenses.

Enhance breeding procedures: Place methods for selective breeding to increase fish growth, survival, and illness resistance.

Make use of cutting-edge technology: To boost production density and water efficiency, consider implementing technologies such as aquaponics or recirculating aquaculture systems (RAS).

8.2 Market Growth

Diversifying your offerings: Expand the range of products you sell by considering cultivating novel species or variants that target certain markets or fetch high prices.

Aim for novel customer segments: Investigate dining establishments, supermarkets, and Internet merchants to broaden your consumer base beyond your typical clientele.

Create channels for direct marketing: To increase your earnings, sell directly to customers via farmers' markets, Internet marketplaces, or their own websites.

Export your seafood: If laws and logistics allow, think about exporting fish to foreign markets where seafood can fetch a better price.

8.3 Addition of Value

Prepare and wrap your fish: Provide products that are smoked, filleted, or have value added to raise the price and perceived value.

Create branded goods: To stand out from competition and fetch higher pricing, give your fish a distinctive brand identity.

Provide supplementary services: To create extra revenue and offer advisory, pond management, or instructional services.

Join forces with other companies: Work together to develop exclusive experiences or package deals with eateries, chefs, or travel companies.

8.4 Extra elements

Sustainability: Incorporate sustainable practices into every aspect of your business to attract eco-aware customers and gain access to exclusive markets.

Marketing and branding: Make an investment in successful marketing and branding techniques to expand the clientele and foster consumer confidence in the offering.

9. Risks and difficulties

1. High Start-Up Cost: The infrastructure, equipment, and fish stocking costs associated with starting a fish hatchery are high.
2. Regulatory Compliance: It imperative to adhere to all applicable local, state, and federal regulations concerning fish health, environmental effects, and water usage. Navigating these regulations is costly and complex.
3. Technical Proficiency: Specialized understanding of fish breeding, feeding, disease prevention, and water quality control is necessary to run successful hatcheries.
4. Market Volatility: Several variables, including the state of the economy, consumer preferences, and competition from alternative fish sources, such as wild fish or other aquaculture operations, might affect the fish market.
5. Disease and Health Risks: If illnesses, parasites, and other health problems are not effectively controlled, fish may suffer severe losses.
6. Impact on the Environment: Negative environmental effects, such as pollution, habitat deterioration, or introduction of exotic species, can be caused by poor management techniques.

7. Temporal Discrepancies: Seasonal variations in production levels could impact revenue streams contingent on the species under cultivation.
8. Market Competition: It difficult to compete with well-established fish hatcheries or other fish production sources, particularly if they have established clientele or economies of scale.
9. Supply Chain Risks: Unexpected events in the supply chain, including feed or equipment shortages, can affect business operations and profitability.
10. Marketing and Distribution: Small-scale hatcheries, in particular, may find it difficult to create efficient marketing plans and distribution networks that will reach their target market.

Conclusion

A fish hatchery can be an excellent way to start a business and significantly boost one's income. Fish raised and sold in a regulated hatchery setting allows business owners to consistently generate a continuous supply of stock. This makes it possible to scale up revenue and production in a way that is challenging to accomplish using traditional fishing alone. Operating a fish hatchery can eventually double or even triple one's revenue with the right financial strategy, investment, and operational execution. However, to be successful, a number of factors need to be carefully considered, including the choice of species, layout of the facility, adherence to laws, and effective marketing. For those prepared to invest the required time, energy, and finances, a successfully managed fish hatchery offers a promising route for substantial financial rewards.

References

Bisht, A., Anand, S., Bhadula, S. and Pal, D. K. (2013). Fish seed production and hatchery management: A Review. New York Science Journal, 6(4):42-48.

Das, S. K. (2003). Breeding of carps using a low-cost, small-scale hatchery in Assam, India: A farmer proven technology. Aquaculture Asia, 8(1):8-10.

Dwivedi, S.N. and Zaidi, S.G.S. (1983). Development of carp hatcheries in India. Fishing chimes. 3(5):29-37.

Gordon, A. S., Plumblee, J., Higdon, G., & Vaughn, D. (2017). Engineering Sustainable Aquaculture in Rural Haiti: A Case Study. International Journal for Service Learning in Engineering, Humanitarian Engineering and Social Entrepreneurship, 12(2), 15-33.

Hossain, M. T., Alam, M. S., Rahman, M. H., Al Asif, A., & Rahmatullah, S. M. (2016). Present status of Indian major carp broodstock management at the hatcheries in Jessore region of Bangladesh. Asian-Australasian Journal of Bioscience and Biotechnology, 1(2), 362-370.

Nasr-Allah, A. M., Dickson, M. W., Al-Kenawy, D. A. R., Ahmed, M. F. M., & El-Naggar, G. O. (2014). Technical characteristics and economic performance of commercial tilapia hatcheries applying different management systems in Egypt. Aquaculture, 426, 222-230.

Piper, R. G. (1982). Fish hatchery management (No. 2175). US Department of the Interior, Fish and Wildlife Service.

12

Optimizing Nutritional Quality in Fried Fish Through Artificial Intelligence Innovation

A. Antony Selvi

Department of Botany, St.Mary's College (Autonomous), Tamil Nadu, India

Abstract

This innovation introduces a system aimed at enhancing the nutritional quality of fried fish through the integration of artificial intelligence. This system consists of three fundamental elements: a microcontroller, a stainless-steel probe featuring a temperature sensor, and an oil-level detection sensor. Additionally, it includes a mobile application powered by artificial intelligence, facilitating the evaluation of the polyunsaturated fatty acid to saturated fatty acid ratio and the index of atherogenicity profile of fried fish. The data collected by the microcontroller undergo thorough processing using an artificial neural network model, along with meta-heuristic optimization techniques such as genetic algorithms and particle swarm optimization (PSO). This processing led to the determination of the PUFA/SFA and IA profiles of fried fish. Upon alignment of the derived data with the predetermined ideal PUFA/SFA and IA values stored in the application database, the application transmits this information to the system, triggering an alert through the integration of a buzzer within the device.

Keywords: *Fried fish, Nutritional quality, Fatty acid profile, Artificial intelligent*

1. Introduction

Food nutrition refers to the energy and nutrients, including proteins, fats, carbohydrates, and more, that the human body derives from food. Nutrition plays a vital role throughout life, serving as the foundation for sustaining bodily functions, promoting growth and development, preventing chronic diseases, enhancing mental well-being, and supporting overall health. According to the 2018 Global Nutrition Report, nutritional imbalances—

whether due to deficiencies or excesses—are a global concern for nutritional security (Fanzo *et al.*, 2018). The "essential" nutrients found in fish include long-chain polyunsaturated fatty acids (LC-PUFAs) such as eicosapentaenoic acid (EPA, 20:5-3), docosahexaenoic acid (DHA, 22:6-3), arachidonic acid (20:4-6), omega-3, and omega-6. As these "essential" fatty acids (FAs) cannot be manufactured by the human body, but have enormous benefits, they must be included in the human diet (Ahmad *et. al.* 2014; Momenzadeh *et. al.*, 2017; Taheri-Garavand *et. al.*, 2018). Fish are frequently prepared using a variety of cooking techniques, such as boiling, baking, roasting, grilling, and frying, to improve their digestibility and flavour and to provide a healthy nutritional profile (Petenuci *et. al.*, 2019; Sadhu *et. al.*, 2020). One of the most popular, simple, and affordable methods of food preparation is the old-fashioned method of frying edible vegetable oil (Lahiri and Ghanta 2008; Lahiri and Ghanta 2010). However, frying at the same time has a negative impact on the nutritional content (Lahiri *et. al.*, 2012; Zhu *et. al.*, 2018). However, frying at the same time has some harmful effects on the nutritional content of food, such as a change in FA profiling that negatively affects nutritional characteristics due to the oxidative breakdown of PUFA (Choudhary *et. al.*, 2014; Kaur *et. al.*, 2019; Şahin and Öztürk 2018). Recent years have seen a sharp rise in the consumption of edible oil for frying food, which poses a serious threat to global health and the economy and is associated with an energy imbalance (Deb, 1995; Domiszewski *et. al.*, 2020; Neff *et. al.*, 2014; Yousefi 2020). The present invention uses artificial intelligence to enhance the nutritional value of fried fish. The content of all referenced publications is hereby integrated to the same extent as if they were specifically and individually cited. If a term's definition in a referenced publication conflicts with the definition provided herein, then the definition prevails. In certain embodiments, quantities or dimensions expressed numerically are understood to be approximate through the use of the term "about." Additionally, the numerical values should be interpreted considering the reported significant digits and standard rounding techniques. Although the numerical ranges and parameters delineating the broad scope of certain embodiments are approximate, values in specific examples are reported with precision to the extent feasible. The ongoing advancement of new sensors and smart wearable devices has accelerated the integration of artificial intelligence (AI) with food nutrition. AI-driven food recommendation systems can offer more tailored dietary suggestions, helping individuals to adopt healthier eating habits and lifestyles. However, creating food recommendations that account for complex factors like taste preferences, sensory differences, cognitive limitations, and physical health is a significant challenge. Yang *et al.* 2021 developed a user interface based on visual tests, along with an online learning framework, to learn users' food preferences and make dietary reco

Fig. 1. Showing the Depicting fried fish being analyzed through artificial intelligence, showcasing nutritional data in a futuristic setting.

Artificial intelligence (AI) is a computerized simulation of human understanding and cognitive processes, enabling machines to perform tasks with human-like decision-making abilities. This technology is valuable for continuous learning and quickly addressing technological errors or issues. AI acts as an innovative tool that mimics human intelligence through the use of computer systems, robotics, and digital devices (Abdallah *et al.*, 2024). AI has numerous applications, such as natural language processing (NLP) to interpret spoken language, computer vision to digitize visual media like videos, speech recognition, and expert systems that replicate human decision-making. The core cognitive functions driving AI encoding are learning (collecting data and creating algorithms to convert it into useful information), reasoning (selecting the right algorithm for a specific goal), and self-correction (continually adjusting algorithms to achieve the most accurate results) (Li *et al.*, 2022). AI is widely used in industries like marketing, finance, healthcare, food quality and safety, food retail, and intelligent process automation. Machine learning (ML), a critical aspect of AI, enhances productivity and fosters innovation (Ben Ayed *et al.*, 2021).

2. Enhancing fried fish preparation through artificial intelligence techniques

Artificial intelligence technology has powerful data processing and analysis capabilities, and it can mine valuable information by establishing the correlation of the data (Figure.2). Therefore, artificial intelligence technology has great potential in predicting and improving the nutritional value of food. Nonetheless, food recommendations must not only align with users' preferences but also take into account their health conditions, as highlighted (Yang *et al.*, 2017)

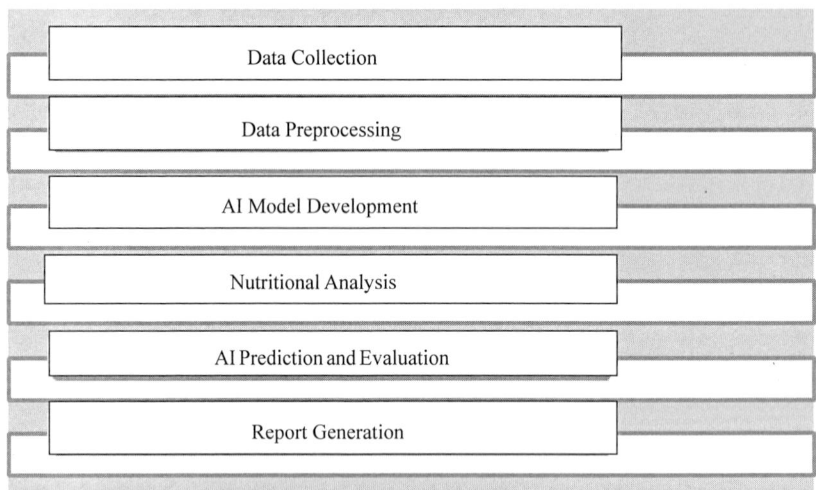

Fig. 2. Shows the Nutritional Value of Fried Fish Using AI Approach

Note: flowchart outlines the process of analyzing the nutritional value of fried fish using an artificial intelligence approach. It involves data collection, processing, and applying AI models for nutrition assessment

Sandhu and colleagues (Sadhu *et al.*, 2022) used a genetic algorithm and particle swarm optimization to predict the nonlinear functional correlation between cooking parameters and the nutritional quality index. Fish is often cooked by different methods including boiling, baking, roasting, grilling, and frying to enhance its digestibility and palatability, with a beneficial nutritional profile (Koubaa *et al.,* 2012). Among all of these, the traditional way of frying with edible vegetable oil is one of the very common, easy, and economic processes for the preparation of food (Oke *et al.,* 2018). By optimizing the combination of cooking parameters, the ratio of polyunsaturated fatty acids (PUFAs) to saturated fatty acids (SFAs) increased to 63.05%, thus enhancing the nutritional value of fried fish. The device described herein utilizes artificial intelligence to boost the nutritional content of fried fish. Figure 1 illustrates the components of the device (200) designed for this purpose. The present invention comprises a microcontroller (150), a temperature sensor (100), a timer (101), an oil level sensor (102), a stainless-steel probe (107), a display (103), Wi-Fi (104), a mobile device (105), a buzzer (108), and an ANN model with stochastic optimization formalisms, such as the genetic algorithm (GA), particle swarm optimization (PSO), and multi-objective genetic algorithm (MOGA) (106). The present invention discloses that the stainless-steel probe (107) is coupled with a temperature sensor (100) and an oil-level sensor (102). The temperature sensor (100) detects the temperature of the oil, and the oil level sensor (102) detects the amount of oil required for frying fish. The present invention also comprises an artificial intelligence-based application that can be installed in a mobile device (105) to determine the polyunsaturated fatty acid (PUFA)/ saturated fatty acids (SFA) ratio and the index of atherogenicity (IA) profile of fried fish. The data received from the microcontroller (150) were processed using an ANN model–based meta-heuristic with stochastic optimization formalisms, genetic algorithm (GA), and particle swarm optimization (PSO), and the polyunsaturated fatty acid (PUFA)/saturated fatty acid (SFA) and index of atherogenicity (IA) profile of fish fry were determined. If the determined data match the ideal PUFA/SFA and IA values pre-stored in the application database, the application sends the information to the device and alerts the user using a buzzer (108) installed in the device. Real-time monitoring and PUFA/SFA and IA values of frying fish can be monitored through the AI-based application interface and on the display (105) of the device. Figure 2 shows the architecture of the proposed invention using the ANN model. After receiving data from the sensors, the microcontroller sends the data to the AI-based mobile application to determine the PUFA/SFA and IA values using ANN modelling. A typical three-layered neural network of MLP, a widely used network in the ANN paradigm, is the input layer (201), hidden layer (202), and output layer (203), which comprises the temperature, time, and oil amount (201). The output layer (203) is the value of the PUFA/SFA ratio and IA. The present invention uses the ANN training procedure, number of nodes in the hidden layer (202), and activation functions used in the hidden and output layers (203). The current disclosure employs automatic testing of all available training techniques, activation functions, and hidden nodes. GA and PSO were used both within and beyond the confines of the inputs to maintain a healthy PUFA/SFA ratio and a low IA value in the diet. In addition to the single-objective functions indicated above, MOGA was used to concurrently achieve all nutritional quality indices with the least amount of oil, temperature, and time to discover a solution that was advantageous to health, cost effectiveness, and time saving. Thus, maintaining essential PUFA-mediated nutritional attributes in fish drastically deteriorated by 44.97% and 99.40% for PUFA/SFA and IA, respectively. In any embodiment described herein, the open-ended terms "comprising," "comprises," and similar terms (synonymous with "including," "having," and "characterized by") may be substituted with the partially closed phrases "consisting essentially of," "consists essentially of," and similar terms, or with the closed phrases "consisting of," "consists of," and similar terms. Additionally, as used herein, the singular forms "a," "an," and "the" encompass both the singular and plural, unless specifically indicated otherwise. Although the embodiments of the present invention have been described with reference to the specific implementations above, it is understood that various changes, modifications, and substitutions can be made. In many cases, certain features of an invention can be utilized independently of others. Additionally, adjustments can be made to the number and arrangement of the components depicted in the figure1 &2.

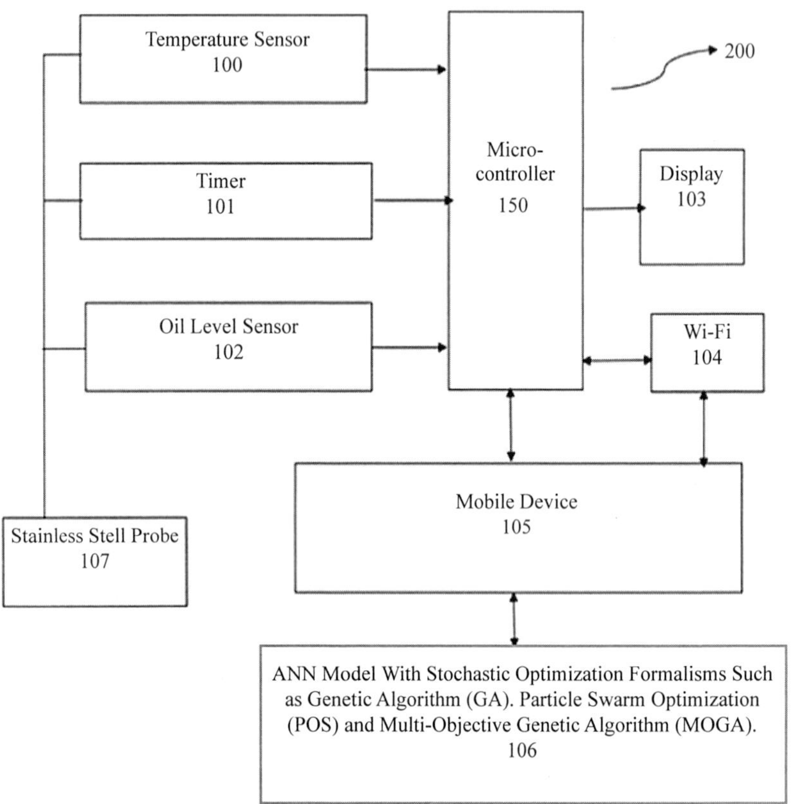

Fig 3. Shows the components of the device used to enhance the nutritional value of fried fish using artificial intelligence

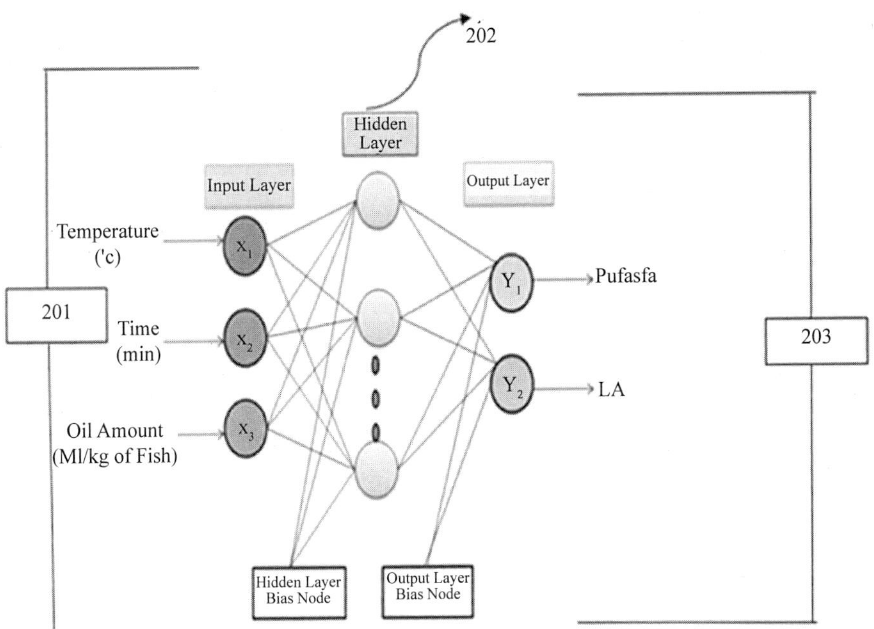

Fig. 4. Implies the architecture of present invention using ANN model

Conclusion

The present invention comprises an artificial intelligence-based application that can be installed in a mobile device to determine polyunsaturated fatty acid (PUFA)/saturated fatty acids (SFA) and the index of atherogenicity (IA) profile of fried fish. The application evaluates the value obtained from the microcontroller by comparing it with the pre-stored values of polyunsaturated fatty acid (PUFA)/saturated fatty acids (SFA) and the index of the atherogenicity (IA) profile. Furthermore, when the nutritional profile of fried fish aligns with the predetermined ideal PUFA/SFA and IA values stored in the application database, the application communicates this information to the device, triggering an alert to the user via a buzzer integrated into the device.

References

Abdallah, R. B., Aribi, H. B., Amara, E. B., Othmeni, I., & Fattouch, S. (2024). Artificial intelligence for sustainable food systems. In B. K. Mishra (Ed.), Fostering cross-industry sustainability with intelligent technologies (pp. 457–15). IGI Global

Ahmad, I., Jeenanunta, C., Chanvarasuth, P., & Komolavanij, S. (2014). Prediction of physical quality parameters of frozen shrimp (litopenaeus vannamei): An artificial neural networks and genetic algorithm approach. Food and Bioprocess Technology, 7(5), 1433–1444. https://doi.org/10.1007/s11947-013-1135-3

Ben Ayed, R., Hanana, M., & Khan, R. (2021). Artificial intelligence to improve the food and agriculture sector. Journal of Food Quality, 2021(open in a new window), 1–7. https://doi.org/10.1155/2021/5584754

Choudhary, M., Sangha, J. K., & Grover, K. (2014). Conventional and nonconventional edible oils: An Indian perspective. Journal of the American Oil Chemists' Society, 91(2), 179–206. https://doi.org/10.1007/s11746-013-2400-3

Deb, K.. (1995). Optimization methods for engineering design. Prentice Hall.

Domiszewski, Z., Duszyńska, K., & Stachowska, E. (2020). Influence of different heat treatments on the lipid quality of African catfish (Clarias Gariepinus). Journal of Aquatic Food Product Technology, 29(9), 886–900. https://doi.org/10.1080/10498850.2020.1817219

Fanzo J., Hawkes C., Udomkesmalee E., Afshin A., Allemandi L., Assery O., Baker P., Battersby J., Bhutta Z., Chen K. (2018). Global Nutrition Report: Shining a Light to Spur Action on Nutrition. Development Initiatives; Bristol, UK: 2018.

Kaur, R., Sharma, A. K., Rani, R., Mawlong, I., & Rai, P. (2019). Medicinal qualities of mustard oil and its role in human health against chronic diseases: A review. Asian Journal of Dairy and Food Research, 38(of), 98–104. https://doi.org/10.18805/ajdfr.DR-1443

Koubaa, A., Mihoubi, N.B., Abdelmouleh, A & Bouain A. (2012) Comparison of the effects of four cooking methods on fatty acid profiles and nutritional composition of red mullet (Mullus barbatus) muscle. Food Sci Biotechnol 21:1243–1250

Lahiri, S. K., & Ghanta, K. C. (2008). Development of an artificial neural network correlation for prediction of hold-up of slurry transport in pipelines. Chemical Engineering Science, 63(6), 1497–1509. https://doi.org/10.1016/j.ces.2007.11.030

Lahiri, S. K., & Ghanta, K. C. (2010). Artificial neural network model with parameter tuning assisted by genetic algorithm technique: Study of critical velocity of slurry flow in pipeline. Asia-Pacific Journal of Chemical Engineering, 5(5), 763–777. https://doi.org/10.1002/apj.403

Lahiri, S. K., Khalfe, N. M., & Wadhwa, S. K. (2012). Particle swarm optimization technique for the optimal design of shell and tube heat exchangers. Chemical Product and Process Modeling, 7(1), 1934–1948. https://doi.org/10.1515/1934-2659.1612

Li, S., Liu, Y., Chen, D., Jiang, Y., Nie, Z., & Pan, F. (2022). Encoding the atomic structure for machine learning in materials science. Wiley Interdisciplinary Reviews: Computational Molecular Science, 12(open in a new window)(1(open in a new window)), e1558. https://doi.org/10.1002/wcms.1558

Momenzadeh, Z., Khodanazary, A., & Ghanemi, K. (2017). Effect of different cooking methods on vitamins, minerals and nutritional quality indices of orange-spotted grouper (Epinephelus coioides). Journal of Food Measurement and Characterization, 11(2), 434–441. https://doi.org/10.1007/s11694-016-9411-3

Neff, M. R., Bhavsar, S. P., Braekevelt, E., & Arts, M. T. (2014). Effects of different cooking methods on fatty acid profiles in four freshwater fishes from the Laurentian Great Lakes region. Food Chemistry, 164, 544–550. https://doi.org/10.1016/j.foodchem.2014.04.104

Oke, E., Idowu, M., Sobukola, O., Adeyeye, S., Akinsola, A. (2018) Frying of food: a critical review. J Culin Sci Technol 16:107–127

Petenuci, M. E., dos Santos, V. J., Gualda, I. P., Lopes, A. P., Schneider, V. V. A., Dos Santos, O. O., & Visentainer, J. V. (2019). Fatty acid composition and nutritional profiles of Brycon spp. from central Amazonia by different methods of quantification. Journal of Food Science and Technology, 56(3), 1551–1558. https://doi.org/10.1007/s13197-019-03654-4

Sadhu T., Banerjee I., Lahiri S.K. & Chakrabarty J. Enhancement of nutritional value of fried fish using an artificial intelligence approach. Environ. Sci. Pollut. Res. 2022;29:20048–20063. doi: 10.1007/s11356-021-13548-8.

Sadhu, T., Banerjee, I., Lahiri, S. K., & Chakrabarty, J. (2020). Modeling and optimization of cooking process parameters to improve the nutritional profile of fried fish by robust hybrid artificial intelligence approach. Journal of Food Process Engineering, 43(9), e13478. https://doi.org/10.1111/jfpe.13478

Şahin, U., & Öztürk, H. K. (2018). Comparison between artificial neural network model and mathematical models for drying kinetics of osmotically dehydrated and fresh figs under open sun drying. Journal of Food Process Engineering, 41(5), e12804. https://doi.org/10.1111/jfpe.12804

Taheri-Garavand, A., Meda, V., & Naderloo, L. (2018). Artificial neural network– genetic algorithm modeling for moisture content prediction of savory leaves drying process in different drying conditions. Engineering in Agriculture, Environment and Food, 11(4), 232–238. https://doi.org/10.1016/j.eaef.2018.08.001

Wang W., Duan L.-Y., Jiang H., Jing P., Song X., Nie L. (2021). Market Dish: Health-aware food recommendation. ACM Trans. Multimed. Comput. Commun. Appl. (TOMM), 17:1–19. doi: 10.1145/3418211.

Yang L., Hsieh C.-K., Yang H., Pollak J.P., Dell N., Belongie S., Cole C., Estrin D. (2017) Yum-me: A personalized nutrient-based meal recommender system. ACM Trans. Inf. Syst. (TOIS) ;36:1–31. doi: 10.1145/3072614.

Zhu, X., Yu, L., Zhou, H., Ma, Q., Zhou, X., Lei, T., Hu, J., Xu, W., Yi, N., & Lei, S. (2018). Atherogenic index of plasma is a novel and better biomarker associated with obesity: A population-based cross-sectional study in China. Lipids in Health and Disease, 17(1), 37. https://doi.org/10.1186/s12944-018-0686-8.

13

Integrated Fish Farming as Future Importance and Increasing Possibilities of Income to Fish Farmers

Shruti Samson and Kamin Alexander

Department of Biological Sciences, Sam Higginbottom University of Agriculture Technology and Sciences, Prayagraj- 211007, Uttar Pradesh, India

Abstract

Integrated fish farming, a sustainable aquaculture practice, combines the cultivation of fish with other complementary activities such as agriculture or livestock farming in a mutually beneficial system. This chapter explores the importance of integrated fish farming as a solution to enhance the income and livelihoods of fish farmers while promoting environmental sustainability (Pauly et al., 2003). By leveraging the synergies between different farming components, integrated fish farming systems can optimize resource utilization, improve productivity, and diversify income streams for farmers. The chapter discusses various integrated fish farming models, including aquaponics, rice-fish culture, and integrated livestock-fish systems, highlighting their potential to increase profitability and resilience in the face of climate change and market uncertainties. Furthermore, the chapter examines the challenges and opportunities associated with integrated fish farming and provides recommendations for policymakers, researchers, and practitioners to promote its widespread adoption and long-term success.

1. Introduction

Integrated fish farming, also known as aqua-agriculture or aquaculture-agriculture integration, represents a sustainable and holistic approach to farming that integrates fish production with other agricultural or livestock activities in a synergistic manner. This integrated approach capitalizes on the complementary relationships between different farming components, resulting in enhanced resource utilization, increased productivity, and improved economic viability for farmers. Unlike traditional monoculture systems, which often rely on single-crop or single-species production, integrated fish farming systems leverage the synergies between diverse farming components to create self-sustaining ecosystems that mimic natural processes and cycles.

Integrated fish farming has gained prominence in recent years as a solution to address multiple challenges facing agriculture and aquaculture sectors. With global population growth, increasing food demand, and dwindling natural resources, there is a growing recognition of the need for more sustainable and resilient food production systems. Integrated fish farming offers a promising alternative by optimizing resource use, diversifying income streams, and promoting environmental sustainability. By combining fish production with other farming activities such as crop cultivation, livestock rearing, or aquaponics, farmers can create multifunctional landscapes that generate multiple benefits for both people and the planet.

2. Importance of Integrated Fish Farming

Integrated fish farming offers numerous advantages that contribute to its importance as a future-oriented farming practice. One of the key benefits is the diversification of income streams for farmers. By integrating fish farming with other agricultural or livestock activities, farmers can reduce their dependence on a single source of revenue and spread their risks across multiple enterprises. This diversification not only improves financial stability but also mitigates risks associated with market fluctuations, climate variability, and environmental uncertainties. Additionally, integrated fish farming systems optimize resource utilization by leveraging the synergies between different farming components. For example, fishponds can be integrated with rice paddies or vegetable gardens, where fish waste serves as fertilizer for crop growth, and plants help maintain water quality by absorbing excess nutrients. This integrated approach not only maximizes the efficiency of inputs but also minimizes waste generation and environmental pollution (Mathias *et al.*, 2020).

Furthermore, integrated fish farming has the potential to increase overall productivity and resilience in farming systems. By combining fish production with other agricultural activities, farmers can create balanced ecosystems that support the growth and development of multiple crops or species. For instance, the presence of fish in rice fields can control pests and weeds, leading to higher rice yields, while the rice canopy provides shade and shelter for fish, promoting their growth and survival (Halwart & Gupta, 2004). Similarly, in aquaponics systems, fish waste provides nutrients for plant growth, while plants help filter and purify the water for fish. This symbiotic relationship results in high yields of both fish and vegetables in a relatively small space. By enhancing productivity and resilience, integrated fish farming systems can contribute to food security, poverty alleviation, and rural development, particularly in small-scale farming communities.

3. Integrated Fish Farming Models

Integrated fish farming encompasses various models and approaches that integrate fish production with other farming activities, each offering unique advantages and opportunities for farmers.

3.1 Aquaponics: Aquaponics represents a closed-loop system where aquaculture and hydroponics are combined synergistically. In this system, fish are raised in tanks, and their waste, rich in nutrients, is utilized to fertilize plants grown hydroponically. The plants, in turn, filter the water, removing waste and purifying it before recirculating it back to the fish tanks. Aquaponics offers several benefits, including high yields of both fish and vegetables, efficient use of water and nutrients, and minimal environmental impact. This model is particularly suitable for urban and peri-urban areas where space is limited, and access to freshwater resources may be constrained (Rakocy et al., 2004).

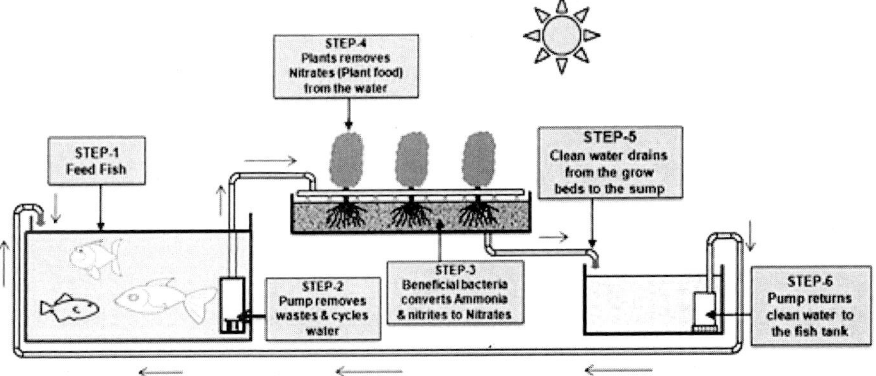

Fig. 1. Aquaponics system In Aquaponics system (Jena *et al.*, 2017).

3.2 Rice-Fish Culture: Rice-fish culture is a traditional farming practice that integrates fish production with rice cultivation in flooded paddies. Fish are stocked in rice fields during the wet season, where they utilize the flooded environment as their habitat. The fish feed on pests and weeds, reducing the need for chemical pesticides and manual labor in weed control. Moreover, fish waste serves as a natural fertilizer for rice plants, enhancing their growth and yield. Rice-fish culture promotes ecological balance in rice ecosystems, improves soil fertility, and provides an additional source of income and protein for farmers (Halwart & Gupta, 2004).

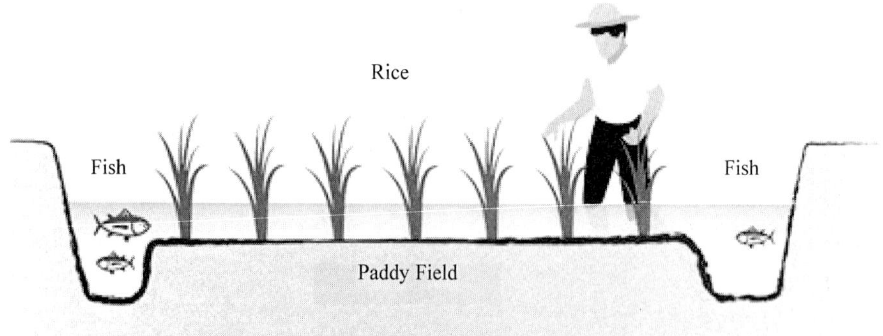

Fig. 2. Rice-Fish Culture (Yi, 2019)

3.3 Integrated Livestock-Fish Systems: Integrated livestock-fish systems combine fish farming with livestock rearing, such as poultry or pigs, in a mutually beneficial manner. Animal wastes, such as manure and leftover feed, are utilized as organic fertilizers for fishponds, supplementing the nutrient requirements of fish and enhancing pond productivity. In return, fishponds serve as a source of supplemental income and protein for livestock farmers. Integrated livestock-fish systems promote nutrient recycling, reduce the environmental impact of livestock farming, and improve overall farm productivity and profitability. Additionally, they offer farmers greater resilience to market fluctuations and input costs by diversifying income streams (Prein *et al.,* 2019).

These integrated fish farming models demonstrate the versatility and adaptability of integrated farming approaches, which can be tailored to suit local conditions, farming practices, and market demands. By harnessing the synergies between different farming components, integrated fish farming systems offer opportunities for sustainable intensification, economic diversification, and environmental stewardship in agricultural landscapes.

4. Challenges and Opportunities

While integrated fish farming offers numerous benefits, it also faces several challenges that need to be addressed for its successful implementation and scaling (Dey *et al.*, 2016).

4.1 Technical Knowledge and Training: Farmers require technical knowledge and training to implement integrated fish farming systems effectively. Capacity-building programs and extension services can play a crucial role in providing farmers with the necessary skills and expertise

4.2 Access to Inputs and Resources: Limited access to inputs such as fish fingerlings, seeds, and equipment can hinder the adoption of integrated fish farming. Improving access to affordable inputs and resources is essential to promote widespread adoption.

4.3 Market Access and Value Chains: Developing robust market linkages and value chains is critical for ensuring that farmers can access markets and obtain fair prices for their products. Supportive policies and infrastructure are needed to facilitate market access and promote value addition.

4.4 Climate Resilience: Integrated fish farming systems offer enhanced resilience to climate change impacts compared to monoculture systems. The diversification of farming activities can buffer against climate-related risks such as droughts, floods, and temperature fluctuations. For example, in areas prone to drought, integrated fish farming can provide alternative income sources and food security during periods of water scarcity by utilizing water-efficient farming practices and diversified cropping systems.

References

Boyd, C. E., & Tucker, C. S. (1998). Pond aquaculture water quality management.

Dey, M. M., Prein, M., & Bosma, R. H. (2016). Integrated rice-fish culture: A well- tested and resource-efficient technology for sustainable food production in Asian wetlands. International Journal of Agricultural Sustainability, 14(3), 262-280.

Halwart, M., & Gupta, M. V. (2004). Culture of fish in rice fields. FAO Fisheries Technical Paper, (498).

Haobijam, J., & Ghosh, S. (2023). Integrated fish farming and its influence on farm livelihoods in Manipur, India. Aquaculture International, 31(4), 2011–2034.

Mathias, J. A., Charles, A. T., & Hu, B. (2020). Integrated Fish Farming.

Pauly, D., Alder, J., Bennett, E., Christensen, V., Tyedmers, P., & Watson, R. (2003). The Future for Fisheries. Science, 302(5649), 1359–1361.

Prein, M., Sharma, R., & Halwart, M. (2019). Integration of aquaculture with crop and livestock farming. In J. E. Flaherty, M. Halwart, & D. L. Bartley (Eds.), The role of aquaculture in sustainable food security (pp. 195-220). Academic Press.

Production. Innovative Farming, 2455-6521. (2)84-89.

Rakocy, J. E., Bailey, D. S., Shultz, R. C., & Thoman, E. S. (2004). Aquaponic production of tilapia and basil: Comparing a batch and staggered cropping system. Aquacultural Engineering, 34(3), 157-164. Springer Science & Business Media.

14

Organic Fish Farming: Certification for Sustainable Income and Future Importance

Kamin Alexander **and** ***Shruti Samson***

Department of Biological Sciences, Sam Higginbottom University of Agriculture, Technology and Sciences, Prayagraj-211007, Uttar Pradesh, India

Abstract

Organic fish farming has emerged as a promising alternative to conventional aquaculture, emphasizing environmentally friendly practices, animal welfare, and product quality. This chapter explores the concept of organic fish farming, its certification process, and its significance for the economic viability of fish farmers. By adhering to organic standards and obtaining certification, fish farmers can access premium markets, command higher prices for their products, and contribute to sustainable food production. The chapter discusses the requirements and benefits of organic certification, challenges faced by organic fish farmers, and the future prospects of organic aquaculture in meeting global food demand while preserving natural resources.

1. Introduction

The global aquaculture industry has experienced rapid growth in recent decades, driven by increasing demand for seafood, declining wild fish stocks, and advancements in aquaculture technology. However, along with this expansion comes a heightened awareness of the environmental and social impacts associated with conventional aquaculture practices. Concerns about water pollution, habitat destruction, and the use of antibiotics and chemicals have led consumers to seek more sustainable alternatives. In response to this demand, organic fish farming has emerged as a viable solution that aligns with principles of environmental conservation, animal welfare, and food quality.

Organic fish farming represents a paradigm shift in aquaculture, emphasizing natural and holistic approaches to production while minimizing negative environmental impacts. Unlike conventional methods that often rely on intensive stocking densities, artificial feeds, and chemical inputs, organic fish farming prioritizes ecological balance, biodiversity preservation, and sustainable resource management. By integrating organic principles into aquaculture operations, farmers can mitigate environmental degradation, enhance ecosystem resilience, and produce seafood that meets stringent organic standards.

In recent years, consumer preferences have shifted towards organic and sustainably produced food products, including seafood. This trend reflects a growing awareness of the links between diet, health, and the environment, as well as concerns about food safety and quality. Organic fish farming offers consumers an assurance of product integrity, free from synthetic chemicals, antibiotics, and genetically modified organisms (GMOs). Furthermore, organic certification provides a means of verifying farming practices, ensuring transparency, and building trust between producers and consumers.

The certification process is a crucial aspect of organic fish farming, providing a framework for adherence to organic standards and verification of compliance. Certification involves rigorous assessment of farming practices, inputs, and management systems by accredited certification bodies, which certify farms based on their adherence to organic principles. Certified organic fish farms must meet criteria related to stocking densities, feed ingredients, water quality management, and disease prevention, among others. Compliance with these standards not only ensures environmental sustainability and product quality but also opens up access to premium markets and higher prices for organic seafood products.

In this chapter, we will explore the principles and practices of organic fish farming, examine the certification process in detail, and discuss the potential economic benefits and future importance of organic aquaculture. Through case studies, research findings, and industry insights, we will highlight the opportunities and challenges associated with organic fish farming and its role in promoting sustainable food production and rural development.

2. The concept of organic fish farming

Organic fish farming embodies a holistic and sustainable approach to aquaculture that prioritizes the health of aquatic ecosystems, animal welfare, and food quality. At its core, organic fish farming seeks to emulate natural processes and minimize environmental impact while producing high-quality seafood. This concept integrates principles of organic agriculture with the unique requirements of fish farming, resulting in a production system that is both environmentally friendly and socially responsible.

One of the fundamental principles of organic fish farming is the use of natural and organic feed ingredients. Unlike conventional aquafeeds, which often contain fishmeal derived from wild- caught fish and synthetic additives, organic feeds are formulated using organic and sustainably sourced ingredients. These may include plant proteins, algae, insects, and by-products from organic food production. By prioritizing organic feed ingredients, organic fish farming reduces reliance on wild fish stocks and minimizes the environmental footprint associated with feed production.

In addition to feed, organic fish farming places a strong emphasis on water quality management and habitat preservation. Farms employ practices such as water recirculation, wetland buffers, and natural filtration systems to maintain clean and oxygen-rich water conditions. By minimizing nutrient runoff, sedimentation, and pollution, organic fish farms help protect surrounding water bodies and support the health of aquatic ecosystems. Furthermore, habitat restoration and enhancement efforts create valuable refuges for native species and contribute to overall biodiversity conservation.

Central to the concept of organic fish farming is the prohibition of synthetic chemicals, antibiotics, and growth promoters. Instead, farms implement natural methods of disease prevention and control, such as biosecurity measures, probiotics, and herbal remedies. By avoiding the use of antibiotics and chemicals, organic fish farming reduces the risk of antibiotic resistance, environmental contamination, and food safety concerns. This focus on natural disease management promotes fish health and resilience, ultimately leading to higher-quality products for consumers.

Animal welfare is another core aspect of organic fish farming, with farms prioritizing the health and well-being of their fish stocks. Farms provide ample space, clean water, and suitable habitat conditions to minimize stress and promote natural behavior. Additionally, organic standards dictate strict stocking densities and handling practices to ensure the welfare of fish throughout their lifecycle. By prioritizing animal welfare, organic fish farming aligns with consumer expectations for ethically produced seafood and fosters trust between producers and consumers.

Certification plays a crucial role in organic fish farming, providing assurance to consumers that products meet specific organic standards and requirements. Farms undergo regular inspections and audits by accredited certification bodies to verify compliance with organic principles and practices. Certified organic products are labeled and identified, allowing consumers to make informed choices about the food they purchase. Furthermore, organic certification ensures transparency and traceability throughout the production chain, from farm to market.

Overall, organic fish farming represents a sustainable and ethical approach to aquaculture that addresses environmental, social, and economic concerns. By embracing organic principles and practices, fish farmers can contribute to the conservation of natural resources, promote animal welfare, and produce seafood that meets the growing demand for sustainable and responsibly sourced food.

3. Benefits of Organic Certification

Organic certification offers numerous benefits for fish farmers, consumers, and the environment:

i) Certified organic fish farmers gain access to premium markets and niche segments of environmentally conscious consumers who prioritize organic, sustainably produced seafood. Organic certification enhances product value, credibility, and marketability, enabling fish farmers to command higher prices for their products and achieve better returns on investment.

ii) Organic fish farming practices promote environmental sustainability by minimizing chemical inputs, reducing pollution, and conserving natural resources. By adopting organic methods, fish farmers contribute to the preservation of aquatic ecosystems, biodiversity, and ecosystem services, such as water purification and habitat provision.

iii) Organic certification provides consumers with assurance that fish products meet stringent organic standards for quality, safety, and sustainability. Certified organic seafood is free from synthetic chemicals, antibiotics, and GMOs, offering consumers a healthier and more environmentally friendly choice.

iv) Organic certification enhances the economic viability and long-term sustainability of fish farming operations by diversifying revenue streams, reducing input costs, and mitigating market risks. Certified organic fish farmers often enjoy greater market demand, price stability, and brand loyalty, leading to increased profitability and resilience in the face of market fluctuations and environmental challenges.

4. Challenges and Opportunities

While organic fish farming offers significant benefits, it also presents challenges and opportunities for fish farmers:

4.1 Transition Period: The transition from conventional to organic fish farming practices may require time, investment, and technical expertise. Fish farmers must undergo a transition period during which they phase out conventional inputs and adopt organic methods, which can pose logistical and financial challenges.

4.2 Cost Considerations: Organic certification may entail additional costs associated with certification fees, compliance requirements, and production inputs. Fish farmers must carefully assess the economic feasibility of organic certification and weigh the potential benefits against the associated costs.

4.3 Technical Knowledge and Training: Implementing organic fish farming practices requires specialized knowledge and skills in areas such as organic feed formulation, disease management, and water quality monitoring. Fish farmers may need access to training programs, extension services, and technical support to successfully transition to organic production methods.

4.4 Regulatory Compliance: Organic fish farming operations are subject to stringent regulatory requirements and organic certification standards. Farmers must ensure compliance with organic regulations, documentation of practices, and adherence to certification criteria to maintain organic status and access organic markets.

4.5 Certification Process: The process of obtaining organic certification can be complex and time-consuming, involving documentation, inspections, and audits by accredited certifying bodies. Fish farmers must navigate the certification process and meet all requirements to achieve and maintain organic certification status.

4.6 Supply Chain Management: Managing the organic supply chain involves coordinating activities across multiple stakeholders, including feed suppliers, hatcheries, processing facilities, and distribution channels. Fish farmers must establish reliable supply chains for organic inputs and ensure traceability and integrity throughout the production chain.

4.7 Consumer Education: Educating consumers about the benefits of organic seafood and the principles of organic farming is essential for building consumer trust and demand. Fish farmers can engage in marketing and outreach efforts to raise awareness about organic aquaculture practices and differentiate their products in the marketplace.

References

Beveridge, M.C.M., Thilsted, S.H., Phillips, & Hall, S.J. (2013). Meeting the food and nutrition needs of the poor: The role of fish and the opportunities and challenges emerging from the rise of aquaculture. Journal of Fish Biology, 83(4), 1067-1084.

Chopin, T., Buschmann, A.H., Halling, C., Troell, M., Kautsky, & Yarish, C. (2001). Integrating seaweeds into marine aquaculture systems: A key toward sustainability. Journal of Phycology, 37(6), 975-986.

Costa-Pierce, B.A. (2008). Ecological Aquaculture: The Evolution of the Blue Revolution. John Wiley & Sons.

Edwards, P. (2014). Aquaculture: Challenges and opportunities. Encyclopedia of Agriculture and Food Systems, 1, 39-51.

FAO. (2018). The State of World Fisheries and Aquaculture 2018 - Meeting the Sustainable Development Goals. Food and Agriculture Organization of the United Nations. Retrieved from http://www.fao.org/3/i9540en/i9540en.pdf

Love, D.C., , E.S., Genello, L., & Semmens, K. (2014). Commercial aquaponics production and profitability: Findings from an international survey. Aquaculture, 435, 67-74.

Troell, M., Naylor, R.L., Metian, M., Beveridge, & Arrow, K.J. (2014). Does aquaculture add resilience to the global food system? Proceedings of the National Academy of Sciences, 111(37), 13257-13263.

15

The Role of Government Policy in Shaping the Future of Fisheries and Aquaculture

Bhooleshwari[1*], Akash Das Vaishnav[3], Narendra Kumar Maurya[1] Dikeshwar Prasad[4] and Khusbu Samal[2]

[1]Department of Aquatic Environment Management, College of Fisheries, Karnataka Veterinary, Animal and Fisheries Sciences University, Mangaluru-575002 Karnataka India

[2]Department of Aquaculture, College of Fisheries, Karnataka Veterinary, Animal and Fisheries Sciences University, Karnataka- 575002 India

[3]Lt. Shri Punaram Nishad college of fisheries, Kawerdha, Dau Shri Vasudev Chandrakar Kamdhenu Vishwavidyalaya University, Durg-491995, Chhattisgarh India

[4]ICAR-Central Institute of Freshwater Aquaculture, Kausalyaganga Bhubaneswar-751002, Odisha, India

Abstract

The global aquaculture and fisheries industries are essential for maintaining environmental sustainability, economic growth, and food security. With these sectors confronting historically unseen obstacles, such as excessive fishing, habitat destruction, and climate change, government policy will play a bigger role in shaping their futures. A summary of the key concepts and elements influencing how government regulations impact the long-term sustainability of fishing and aquaculture industries. The union administration is committed to providing a favorable climate in collaboration with the States and UTs so that the fishing industry can realize its full potential, achieve excellence, and fulfill its obligations to farmers and customers. To achieve the policy objectives, the government will enlist the support and involvement of all relevant parties. The National Fisheries Policy 2020 provides a planned path ahead for responsible and sustainable development, exploitation, management, and regulation of capture and culture fisheries. In order to achieve the objectives of the "Blue Economy," the Policy will guarantee fruitful integration with other economic sectors, including agriculture, coastal area development, and eco-tourism. Universities are also involved in accessing resources and knowledge.

Keywords: *National Fisheries Policy, PMMSY, Schemes of Fisheries Sector, Strategies of Government. policies.*

1. Introduction

The Ministry of Fisheries, Animal Husbandry, and Dairying is principally responsible for overseeing the fisheries policy in India. The development of infrastructure and facilities for the fishing sector, ensuring the well-being of fishers, encouraging the growth of the aquaculture sector, and environmentally friendly growth and administration of fish resources are among the primary goals of these policies. The government also administers schemes such as the Pradhan Mantri Matsya Sampada Yojana (PMMSY) to promote sustainable fishing techniques and enhance the fish production in the country. The government also grants fishers the opportunity to purchase boats and fishing equipment, as well as to build ice manufacturing facilities and fish landing hubs.

2. Brief about Aquatic Resources and Fish Production

India possesses an abundance of varied and abundant fishery resources, encompassing deep seas, lakes, ponds, rivers, and over 10% of the world's fish and shellfish species. The wide coastline of the nation, the 2.02 million square km Exclusive Economic Zone (EEZ), and the 0.53 million square km continental shelf area are all home to the nation's marine fisheries resources. The freshwater resources consist of 1.95 lakh km of rivers and canals, 8.12 lakh hectares of lakes in floodplains, 24.1 lakh hectares of ponds and tanks, 31.5 lakh hectares of reservoirs, 12.4 lakh hectares of brackish water, 12 lakh hectares of saline/alkaline impacted areas, etc. The huge and diversified inland resources that are currently untapped present excellent prospects for generating income and bringing about economic success (Handbook of Statistics on Indian State, RBI, 2021). The total amount of marine fish landings along the mainland of India in 2022 was estimated to be 3.49 million tons, a 14.53% increase from 2021. Fish landings increased significantly in the year by 28.02% compared to 2020, the year that the pandemic affected. The forecast for 2022 was 2.0% less than the pre-COVID year of 2019, notwithstanding these advancements (Yan *et al.*, 2021). With 7.22 lakh tonnes of fish landings, Tamil Nadu topped the list of coastal states. Karnataka came in second with 6.95 lakh tonnes, while Kerala came in third with 6.87 lakh tonnes. With 5.03 lakh tonnes, Gujarat, which had previously held the top spots, fell to the fourth place. Tamil Nadu, Karnataka, Kerala, and Gujarat made up 20.69%, 19.90%, 19.68%, and 14.40% of the country's total states. The relative shares of these four states - Tamil Nadu, Karnataka, Kerala, and Gujarat - in the national total were 20.69%, 19.90%,

19.68%, and 14.40%. Fish landings have increased in every state except for Gujarat and Odisha as compared to 2021(CMFRI Annual Report 2022). The Indian fisheries industry, which is regarded as the economy's "developing star," has evolved from its historical status as a supplemental or poverty industry to a significant one. More than 16 million primary fishermen depend on the industry for their living; throughout the supply chain, the figure almost doubles (National Fisheries Policy 2020).

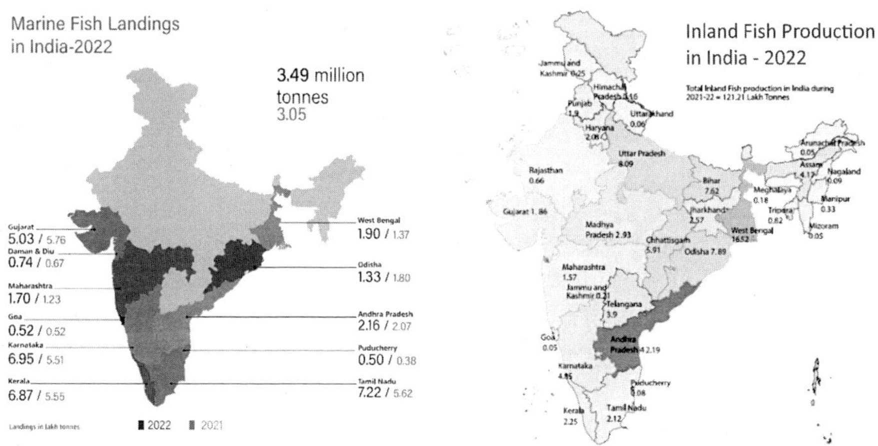

Fig. Fish production in India in the year of 2022.
Source: CMFRI Annual Report 2022, Handbook of Statistics on Indian State -RBI (2021)

3. Need for Government Policy

An estimated 28 million people, or roughly 2.04% of the country's population, live as fishers and fish farmers (Lakra & Krishnan, (2012).). The industry presents enormous potential for creating jobs, particularly for women and young people, generating foreign exchange earnings, and ensuring the country's food and nutritional security. India's fisheries sector has great potential for expanding its reach to historically untapped or underutilized resources in both inland and marine waters, significantly increasing aquaculture productivity and production, integrating farming productively with other industries such as agriculture, horticulture, poultry, and livestock, expanding non-food fisheries such as ornamental fisheries, and improving the availability of omega-3 fatty acid-rich fish protein for the country's expanding population. In general, ponds, tanks, and floodplains may all be very helpful in collecting and storing rainwater, while also acting as a valuable ameliorating agent to help replenish water from the ground.

The necessity of environmentally conscious resource development is now more important than ever because of the nation's expanding population and rising need for seafood protein. The nation must create a strong framework for its national fisheries policy to meet these demands and guarantee a growth trajectory that meets present needs and leaves space for an even stronger fishery in the future. The Policy will offer a guide on how to best utilize the resources of catch and culture fisheries to maintain the targeted levels of productivity and production. It is anticipated that in the near future, this policy framework will direct activities of a similar nature at the State and Union Territory levels (Rajeev & Bhandarkar, 2022).

3.1 What is Policy?

We describe policies that define the expected behavior of distributed heterogeneous systems, applications, and networks, and derive them from management goals (Wiles, 1994). Policies are evolving concepts. Policies have the authority to initiate or change the current management processes. Policies can be used to effectively govern distributed heterogeneous systems, networks, and applications. Lastly, the policy reflects national aspirations and the development goals set by the nation's leadership, taking inspiration from the fact that the government has established a separate ministry for the fisheries sector. This will ensure that fisheries become an equal partner with other developmental sectors to make India a USD 5.0 trillion economy by the year 2025.

3.1.1 Vision: "A healthy and vibrant fisheries sector that meets the needs of the present and future generations."

3.1.2 Mission: "While keeping the sustainability of the resources at the core of all actions, the National Fisheries Policy will meet the social and economic goals and well-being of the fishers and fish farmers and is intended to guide the coordination and management of the fisheries sector in the country during the next ten years."

3.1.3 Objectives: Securing the nation's capture fisheries and the overall development of the aquaculture industry is the goal of the National Fisheries Policy. The primary focus of the Policy will be on fishers and fish farmers, but it will also aim to protect and develop the resources and their associated habitats in a sustainable manner, preserve the integrity of the ecosystem, meet the growing population's needs for food and nutrition, uphold the rights of fishing and farming communities and foster their resilience, increase the competitiveness of Indian fish and fish products and wise use of fisheries resources.

3.2 The following are the goals of the National Fisheries & Aquaculture Policy (NFAP):

1. Boost small- and large-scale aquaculture production.
2. Annual fish yield from the capture fisheries.
3. Fortify participatory fisheries management regimes.
4. Goals Reduce post-harvest losses by half.
5. Annual boost fish exports.
6. Put the stability of resources at the center of modern, rationalized (i.e., science and technology), and diverse fishing methods in oceans and seas.
7. Encourage inland fishing and aquaculture using standardized methods of production, inputs, and cultivation to support sustainable and ethical inland fishing and culture.
8. Maintain and protect native fish populations, their habitats, and the ecology that supports them.
9. Globally competitive and value-added fish products that meet international standards are marketed, traded, and exported.
10. Encouraging successful collaborative action, private involvement, and community collaboration in the fishing industry.

3.3 Fisheries Management Structure in India

Since fisheries are a state issue, state legislation is essential for the control of fisheries. According to the principles of cooperative federalism, the Central Government is accountable for enhancing the efforts of the former in this domain. While the Central and Coastal State/UT Governments share responsibility for managing marine fisheries, State Governments are solely responsible for inland fisheries (Gangal *et al.*, 2023). Coastal states and territories manage fisheries in marine waters within the jurisdictional boundary of 12 nautical miles (22 km). The Indian government is in charge of developing, managing, and controlling fisheries in the Exclusive Economic Zone (EEZ) up to 200 nautical miles (370 km) and beyond under the jurisdiction of the Government of India.

3.4 Constraints in the Growth of Fisheries Sector

Major constraints and some of the other factors constraining the growth of capture fisheries impacting the growth of marine capture fisheries include limited scope for expansion due to:

- Overcapacity in territorial waters, weak regulations, inefficient management, and prevalence of traditional fishing practices.

- Inadequate infrastructure, especially fishing harbors, landing centers, cold chain and distribution systems, poor processing and value addition, waste, traceability, and certification.
- Non-availability of skilled personnel.
- Depleted stocks in natural waters.
- Issues Related to Tenure and Leased Rights.
- The use of obsolete technology for harvesting, coupled with low capital infusion, is a significant limiting factor.

3.5 Issues that are impeding the expansion of culture fisheries include

- Inadequate overall condition of the supplies, in particular the quantity and quality of water
- Cultivation system with minimal input
- Reduced production due to a lack of variety in species and cultural activities
- Insufficient framework for regulation
- Increasing rates of illness (disease)
- Minimal investment levels
- High credit costs and limited availability of institutional credit
- Insufficient facilities for production and pre-production
- Facilities for post-harvest and processing
- Limited adoption of technologies and dearth of competent staff in aquaculture and extended services.

3.6 India's Principal Fisheries Policies

Regarding the regulation of access to the usage of publicly accessible open waterbodies for aquaculture, including cage and pen culture, national and local policies vary throughout nations. Along with pond culture and other techniques, investing in cage culture in publicly accessible open water bodies has been shown to be a successful and efficient way to boost aquaculture productivity when properly regulated.

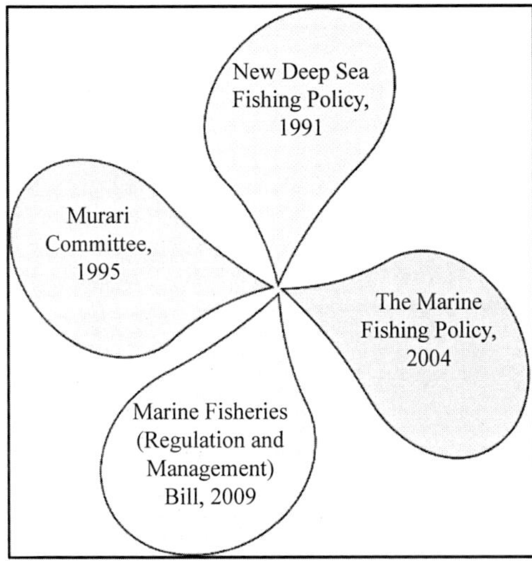

Fig. 1. The Indian fisheries policies and committees

Table 1. Policies and their plans (National Fisheries Policy 2020).

S.No.	Policies	Brief about Policies	Plans
1.	New Deep Sea Fishing Policy, 1991	The Indian government introduced the New Deep Sea Fishing Policy (NDSP), a part of the strategy for economic transformation.	1. Forming 49:51 equity joint ventures in deep sea fishing, processing, and marketing between international and Indian businesses. 2. Letting foreign fishing boats use Indian EEZ for operations. 3. Testing foreign fishing vessels through engagement.
2.	Murari Committee, 1995	The Murari Committee was formed and now has 41 members, including experts, officials and delegates from fishing villages. It was divided into five groups and toured all of the states with coastlines to get input from different fishing sectors.	1. In the future, there won't be any joint venture, charter, lease, or test fishing vessel permits renewed, extended, or granted new licenses. 2. In compliance with the relevant legal procedures, the existing licenses will be revoked. 3. Boost fishing community knowledge to facilitate the utilization of deep-sea resources and the avoidance of pollution. 4. Fuel availability at a discount 5. Fishing restrictions across the whole EEZ 6. A distinct ministry to manage the fisheries in its entirety.

3.	The Marine Fishing Policy, 2004	Most deep-sea resources are located outside the traditional fishing boundaries and the reach of native boats. This can only be used profitably if fishing vessels that are appropriately sized and capable are introduced into the fishery.	1. The rule also prohibits the capture of non-targeted species including juveniles. 2. Monitoring, control, and surveillance, or MCS, would be put into practice by putting observers on board commercial fishing vessels operating in the Indian EEZ. 3. Development of Fisheries in Lakshadweep and Andaman & Nicobar Islands Union Territories
4.	Marine Fisheries (Regulation and Management) Bill, 2009	It establishes a common legal framework for the regulation of fisheries and the conservation and sustainable use of fishery resources in all maritime zones, including territorial waters.	1. The EEZ may extend up to 200 nautical miles from the base line, the contiguous zone may be up to 24 nautical miles from the base line, and the territorial waters may be up to 12 nautical miles from the base line. 2. It proposes to cover Indian fishing vessels manufactured in India, owners of such vessels, the crew members working in the fisheries and fishing on board these vessels, and their operations, notably in the EEZ.

India ranks third globally in fish production. The Indian government has been implementing programs such as the Blue Revolution Scheme and Pradhan Mantri Matsya Sampada Yojana (PMMSY) to increase the nation's fish supply.

3.7 Proposals to Boost India's Fish Production by 2023

The Indian government has implemented several schemes to increase the nation's total fish output and efficiency. These initiatives also aim to provide food security and create jobs. Among the most popular programs managed by the government are

1. Pradhan Mantri Matsya Sampada Yojana (PMMSY)
2. Fisheries and Aquaculture Infrastructure Development Fund (FIDF)
3. Blue Revolution Scheme
4. Fisheries Kisan Credit Card Scheme

3.8 Other Schemes related to Fishing Sector

1. Palk Bay Scheme
2. Integrated Fisheries Development Scheme (IFDS)

The country's entire territory as well as its Exclusive Economic Zone (EEZ) are covered by the National Fisheries Policy (NFP), which has a ten-year timeline (2021–2030) (National Fisheries Policy 2020).

4. The following explanation highlights the fundamental principles of government policy pertaining to strategy:

4.1 Marine Fisheries

Early in the advancement strategy process in India, the potential of the fisheries industry in general and the marine fisheries subsector in particular was acknowledged. Marine fisheries serve as a vital source of food and nourishment for the populace, but they also contribute significantly to trade and commerce, which in turn support maritime populations' lifestyles and opportunities for employment.

4.1.1 Sustaining marine fisheries

The paramount requirement to ensure sustainability (ecological, economic, social, and institutional) would be to rationalize fishing efforts in relation to environmental stability.

4.1.2 The following are the main areas where quick action is required.

1. Encourage investments in harvest and post-harvest facilities and deep-sea fishing in Areas Beyond National Jurisdiction (ABNJ) to unlock underutilized resources through the use of appropriate and responsible technology and capacity building for stakeholders, particularly artisanal fishers.
2. Enhancing fishing efforts and creating and implementing management strategies to restore collapsed or deteriorated fish supplies.
3. Establishing a comprehensive strategy in consultation for resource exploitation in the EEZ and developing conservation measures through consultative methods, such as species-specific and zonal/area-specific management strategies.
4. Encouraging the use of co-management and EAFM techniques.
5. Encouraging coastal states and units to expand the area set aside under MFRAs for non-mechanized fishing boat operators.
6. Enabling knowledge management throughout the marine fishing industry through the application of cutting-edge IT technology.
7. Establishing deep-sea and offshore fishing while maintaining the national security and safety of fishermen

4.1.3 Observation, Regulation and Supervision

As a signatory to numerous international agreements and arrangements aimed at discouraging, preventing, and outlawing illicit, unreported, and unregulated (IUU) fishing, the Indian government will set up a reliable system at the port and sea to guarantee that its fishing fleet stays out of the ABNJ and its own Exclusive Economic Zone (EEZ).

4.1.4 The following are the main areas that require immediate intervention

1. Establishing an efficient MCS system in accordance with the NPOA-MCS, in collaboration with coastal states/UTs and other relevant ministries/departments. Along with strengthening the community's role in carrying out MCS functions, this will also entail improving the abilities and talent of MCS managers.
2. Encouraging national and international fishing vessels to take action against IUU fishing.

4.2 Inland Fisheries

The resources available for freshwater fisheries that are captured are just as abundant and diverse as those available for marine fisheries, and they have always been just as important to the population's ability to survive and grow food and nourishment. Although their movement to other means of income is more noticeable than that of any other sector involved in food production, riparian communities along India's major river systems are just as old and traditional as marine fishers. This is because of the shifting inland sector. The resources for inland capture fisheries consist of 1.2 million hectares of floodplains, 3.52 million ha of small and large reservoirs, and a riverine length of 2,01,496 km (including tributaries and irrigational canals). When rivers and canals were removed from the equation, the projected total area available for inland fishing was 8.24 million ha. It is estimated that there are 24.29 million inland fishers worldwide. overseeing fisheries in the wetlands, natural lakes, and floodplains of Indian rivers (Handbook of Statistics on Indian State, RBI, 2021).

4.2.1 The following are the main areas that require quick attention

1. Guaranteeing that there is sufficient flow of water in rivers and their tributaries to support fisheries.
2. The ecological well-being of the riverine environment can be enhanced by reducing pollution entering rivers and their tributaries from both point and non-point sources in collaboration with pertinent agencies.

3. Safeguarding the riverine sections and corresponding floodplains to guarantee the survival of the endemic species population by establishing protected areas, implementing time and area closures, and managing efforts.
4. To revitalize these resources and exploit their numerous ecosystem services, the link between rivers and floodplains must be restored.
5. Leasing laws must be put in place to guarantee that resources are used according to their productivity, and to provide local communities with the authority to manage resources.
6. Establishing the infrastructure required for stocking and seed production.

4.3 Brackish Water Aquaculture

The Coastal Aquaculture Authority Act, 2005, which is required by law to guarantee environmentally friendly farming methods that do not negatively impact the natural resources and ecosystems of the country's shoreline regions, supports coastal aquaculture, including brackish water shrimp farming and other aquaculture activities. The Act's established Rules and Guidelines provide the industry with the direction it needs to establish and run shrimp farms.

4.3.1 Immediate assistance is necessary in the following critical areas

1. Encouraging species diversity by bringing back P. indicus and tiger shrimp, and adding finfish such as mullets and seabass.
2. Encouraging the domestication of P. indicus and tiger shrimp and working toward the creation of a particular pathogen-free germplasm.
3. Encouraging more shrimp farming to be established to effectively use coastal areas, generate jobs, and improve food and nutritional security.
4. Developing biosecurity and health management, completely abstaining from the use of prohibited antibiotics, building more quarantine facilities, and promoting BMPs and GAP.
5. Establishing and running shrimp farms.

4.4 Mariculture

The Indian government has started to develop mariculture in the nation's coastal regions after realizing that the country's need for seafood would only grow with time and that the country's marine capture fisheries might not be able to supply all the additional demand. The country's prospective areas for mariculture growth have led to projections of annual yield of 4-8 mmt. The policy measures would be numerous, beginning with the creation of a blueprint of appropriate sites/areas throughout the Indian coastline with a leasing policy based on a Marine Spatial Planning (MSP) approach, given that mariculture growth in the nation is still in its infancy.

4.4.1 The following are the main areas that require immediate intervention

1. Creating a development plan for mariculture that considers suitable species, areas both inside and outside territorial waters, leasing policies, and support services in collaboration with coastal States and UTs.
2. To address the needs of local fishermen, research and development for cage fabrication, which is cost-effective and based on local technology, is supported.
3. Establishing brood banks and commercial hatcheries for seed production are encouraged.

4.5 Seaweed Farming

Seaweed farming in the nation has unrealized potential, similar to aquaculture and mariculture. The demand for seaweed is rising worldwide because of its potential use as a source of biofuel, hydrocolloids, cosmetics, and nutritional supplements. This subsector can have a major impact on the economy of the nation's coastal regions and offers vast scope for value creation along the value chain. Assisting in the R&D of cost-effective, locally based cage construction to satisfy the needs of the local fishing community. India is home to approximately 844 different species of seaweed, with a standing stock of 58,715 metric tons. Seaweeds are widely distributed across the Andaman & Nicobar Islands, Lakshadweep, and the shores of Gujarat, Diu, and Tamil Nadu. In Tamil Nadu, Andhra Pradesh, Orissa, and surrounding Mumbai, Ratnagiri, Goa, Karwar, Varkala, Vizhinjam, and Pulicat, there are rich seaweed beds. However, natural seaweed resources have been depleted as a result of ongoing, haphazard, and uncontrolled collections (Handbook of Statistics on Indian State, RBI, 2021).

4.5.1 The following are the main areas where quick action is required

1. Encouraging the cultivation of seaweed in offshore waters, landward and seaward along the shoreline, as well as in integrated multitrophic aquaculture (IMTA) systems.
2. Seaweed seed banks were established along the coast to supply the basic materials needed to cultivate seaweed.
3. Provide the resources required to draw fishermen to seaweed farms.
4. Encouraging women from coastal communities, including fisherwomen, to pursue seaweed cultivation.

5. Encouraging business owners to establish small- and medium-sized processing facilities to produce goods using seaweed.
6. Enabling product-marketing to stimulate the sector.

4.6 Ornamental fish farming

Fish used for ornamentation comprises a small but significant portion of the global fish trade. Maintaining an aquarium is a common hobby, and the annual global trade in ornamental fish is estimated to be worth US$18–20 billion. India has an abundance of commercial and valuable ornamental fish resources owing to its various biological resources. More than 300 sea species and approximately 374 freshwater species are suitable for ornamental fisheries. The policy would support the establishment of domestic facilities to increase ornamental fish production, the creation of an end-to-end supply chain, and habitat preservation to improve rural livelihoods. In addition, the policy would support the building of aquariums in public spaces and consumer education regarding the advantages of aquarium maintenance. Using proper interventions, women and youth will receive special attention (National Fisheries Policy 2020).

4.6.1 One of the main areas that need immediate attention is:

1. Establishment of an ornamental supply chain from beginning to end.
2. Enabling procurement of breeding stock overseas.
3. Encouraging commercial-scale operators to produce and distribute larvae to homes that can raise them in exchange for buyback agreements.
4. Encouraging aquariums in residences and workplaces.
5. Developing entrepreneurship among rural women and youth by assisting them in establishing small- to medium-sized decorative businesses.
6. Helping small and medium-sized businesses construct aquariums and additional accessories to foster the growth of ornamental fisheries.

4.7 Post-harvest & Trade

All entities involved in the synchronized production and value-adding processes required to create food items comprise a food value chain or FVC. The essential characteristics of a sustainable FVC guarantee that it is

i) It is financially successful at every stage (economic sustainability)
ii) It helps society broadly (social sustainability)
iii) It has a neutral or favorable effect on the environment (environmental sustainability).

4.7.1 The following actions are the policy targets to achieve these goals

4.7.1.1 Enhancing value and supply chain

- Improving distribution methods to achieve zero waste is one of the main areas that require quick intervention.
- Developing the abilities of value chain players to add value and create value, and enabling and encouraging them to do so.
- Encourage the eco-labeling of certain fisheries while involving all relevant parties.

4.7.1.2 Domestic marketing development

- Developing consumer knowledge about the advantages of eating fish and investigating new approaches to increase fish consumption.
- Enhancing supply networks to enable fish accessibility, affordability, and availability.
- Encouraging the creation of new products and innovative marketing strategies, such as Internet marketing.
- Encouraging tightly knit regional producer-processor value chains to lessen the need for intermediaries.

4.7.1.3 Encouraging participants in commerce and food safety

- Encouraging diversification of products and species to increase export value and market share.
- Investigating novel markets and endorsing seafood branded "India."
- Enhancing cleanliness and hygiene in fishhouses and fish processing facilities to guarantee fish and fish products that fulfill global standards.
- Encouraging the use of conservation technologies to protect endangered species.
- Including traceability throughout the supply chain
- Encouraging tightly knit regional producer-processor value chains to lessen the need for intermediaries.

4.8 Fisheries Governance

- Controlling prudent and environmentally friendly utilization of marine and terrestrial resources.
- Entrepreneurship and HRD
- Database
- Imparting expertise and providing technical assistance

4.9 Regional/International Commitments

- Promoting collaboration between regions.
- Establishing India as a regional leader.

4.9.1 Common Thread

1. Coordination between sectors
2. Utilization and Handling of Water
3. Water body leasing, licensing and management for culture fisheries
4. Aquaculture at the Level of Agriculture
5. A Cluster Approach to Aquaculture Development
6. Zoning for aquaculture and spatial planning
7. Disease monitoring and health management
8. Verification and Trackability
9. Diversification
10. Controlling exotic materials
11. Development of Genetic Upgradation Research and Fish Seed Production
12. Post-Harvest and Value-Added
13. Trade and Food Safety
14. Extension and Assistance Framework
15. Enhancement of Human Resources
16. Involvement of the Society, Cooperatives, and Farmers' Organizations
17. Welfare and Parity by Gender
18. Climate Change, Biodiversity, Ecosystems, and Disaster Management
19. Organizational funding, individual investment, and public–private collaboration.

4.10 Government policy outcomes in determining how fisheries and aquaculture develop in the future

It is anticipated that by 2030, the following will occur:

- Simplifying technology introduction with the purpose of enhancing culture systems, breeding, seed raising, feed production, and disease control.
- Accessibility of a network of facilities for monitoring, controlling operations, and data mining to ascertain the state of different resources and activities, spanning from the field to the central level.
- Integration of scientific knowledge to improve culture systems, breeding, seed raising, feed production, and disease control.

- Reinforced breeding and proliferation centers, feed mills, hatcheries, nurseries, and quarantine facilities are among the key components of production infrastructure.
- The market drives operations to address price signals and resource constraints in domestic and international markets.
- The establishment of supplementary post-harvest capabilities via cold chain facilities, processing, and value addition will result in a decrease in waste to less than 10%, which will raise farmers' income.
- Up to one lakh crore worth of new projects, such as clusters, zoning, certification, labelling, and branding, would increase the acceptability of products for export.
- Granting fishermen, farmers, and fish workers the freedom to decide what agricultural species, practices, and food to sell to increase revenue and improve living conditions.
- Over the next ten years, there will be a six-fold increase in investment in the fishing sector. (Department of Fisheries, 2020)

Conclusion

In conclusion, the development and execution of successful governmental policies are essential for the future of aquaculture and fisheries. To guarantee the sustainability and resilience of these vital sectors in the face of new challenges, a comprehensive strategy that addresses the environmental, social, and economic dimensions is required. Governments will always play a major role in the establishment of comprehensive and flexible policy frameworks; therefore, cooperation between policymakers, industrial stakeholders, and the scientific community is essential.

References

CMFRI Annual Report (2022), Central Marine Fisheries Research Institute, Kochi. 252.

Department of Fisheries (2024, January 18) Department of Fisheries. Department of fisheries ministry of fisheries, animal husbandry and dairying government of india. https://www.bing.com/ck/a.

Department of Fisheries. (2020) Annual Report 2019 - 20. Ministry of Fisheries, Animal Husbandry and Dairying, Government of India. http://dof.gov.in/sites/default/files/2020-10/Annual_Report.

Gangal, M., Suri, V., & Arthur, R. (2023) How well does Indian fisheries policy engage with fisheries biology? Exploring the science-policy interface of coastal capture fisheries along the west coast of India. Marine Policy, 156:105796. https://doi.org/10.1016/j.marpol.2023.105796.

GoI (2020). National Fisheries Policy 2020 Draft, Department of Fisheries, Government of India.

Handbook of Statistics on Indian States, (2021-22) Director, Division of Reports and Knowledge Dissemination, Department of Economic and Policy Research (DRKD, DEPR), Reserve Bank of India, Amar Building, Ground Floor,

Sir P. M. Road, Fort, P. B. No.1036, Mumbai - 400 001.

Lakra, W. S., & Krishnan, M. (2012). Livelihoods, Employment and Income for Marginal Fishers and Fish Farmers in India. Publication division directorate of economics and statistics department of agriculture and co-operation ministry of agriculture government of India, 513.

Rajeev, M., & Bhandarkar, S. (2022). Fisheries Sector in India - An Overview. Unravelling Supply Chain Networks of Fisheries in India: The Transformation of Retail, 47-59. https://doi.org/10.1007/978-981-16-7603-1_4.

Wies, R. (1994) Policies in network and systems management—Formal definition and architecture. J. Netw. Syst. Manag., 2:63-83. https://doi.org/10.1007/BF02141605.

Yan He, B., F., Yu, H., Su, F., Lyne, V., Cui, Y., Kang, L. and Wu, W., (2021). Global fisheries responses to culture, policy and COVID-19 from 2017 to 2020. Remote Sensing, 13(22):4507. https://doi.org/10.3390/rs13224507.

16

Aquaculture Finance and Rural Development

Sagar Samanta[1], Prabir Sahoo[2*] and Joydeep Maity[2]

[1]Department of Fisheries Science, Medinipur City College
West Medinipur-721 129, West Bengal, India

[2]Department of Coastal Aquaculture, Centre of Advanced Study in Marine Sciences, Annamalai University, Cuddalore-608 502, Tamil Nadu, India

Abstract

Rural development is greatly aided by aquaculture funding, which improves rural people's standards of living and promotes sustainable economic growth. By analyzing its effects on job possibilities, food security, income production, and poverty reduction, this study investigates the role of aquaculture finance in rural development. It addresses several funding sources and tools used in aquaculture, such as grants, microloans, investment plans, and microfinance, emphasizing the success of these financial tools in helping small-scale rural farmers and business owners. This study also discusses market linkages, risk management, regulatory frameworks, credit availability, and other issues related to aquaculture finance. Legislators, financial institutions, and other stakeholders can create plans to optimize the socioeconomic advantages of aquaculture while fostering inclusive growth by comprehending the relationship between financing for aquaculture and rural development.

Keywords: *Aquaculture, Finance, Rural Development, Economic Growth, Income Generation, Challenges, Opportunities, Sustainability.*

1. Introduction

Growing at an average annual rate of 6.2 percent between 2006 and 2012, aquaculture is one of the food-producing industries with the fastest rates of growth in the world (FAO, 2014). Aquaculture production has more than doubled globally, reaching 185 million tons (FAO, 2023). Aquaculture will need to be expanded to feed almost three billion people who depend on fish and shellfish as their main sources of protein, while wild capture fisheries are expected to level off in the upcoming years (Tveteras *et al.*, 6 2012). More than 93.5 percent of the predicted 59.5 million persons working in aquaculture production worldwide are employed in Asia (FAO, 2022).

There are many aspects of the process of improving the well-being of rural men, women, and children, and of the rural economy growing steadily; however, the growth of the agricultural sector in particular is thought to be the primary driver of both ensuring food security for all and decreasing poverty and hunger. The rural poor can only become more food secure if aquaculture grows faster in nations with rural poverty. This will increase both farm and nonfarm income in rural areas. Diverse forms of aquaculture play significant roles in the development of farming and agricultural systems. Providing food with a high nutritional value, creating income and jobs, lowering the risk of monoculture production failure, improving access to water, improving aquatic resource management, and boosting farm sustainability can help reduce food insecurity, malnutrition, and poverty (Prein and Ahmed, 2000). Furthermore, farmers depend on other value chain participants to establish prices and sell their goods, because they have weak market connections and little market knowledge. Because of this disparity in power, value chain participants are more likely to engage in corrupt practices, which frequently results in smallholders (the original producers) being offered a discounted market price.

There are 908 million people living in rural areas of India (Rathore, 2024). The absence of sufficient basic services, low productivity, and poverty are hallmarks of rural communities. Since independence, one of the main priorities of development planning has been the development of rural areas and the significance of the social and economic situation of people living in rural communities. Realizing that the benefits of progress do not always trickle down to the most marginalized groups in society, the government has turned to active involvement in targeted redistributive policies. The Government of the Countryside and Employment Ministry of India, in addition to building infrastructure, has been creating targeted programs that include various agricultural methods to reduce poverty in rural areas.

2. Historical Background

2.1 In the early 1950s and the 1960s

Introduction: During the 1950s and the 1960s, the idea of aquaculture began to garner traction in India, mainly as a way to boost fish output and solve issues related to food security.

Government initiatives: The Indian government established fish farms and training facilities as part of several measures to support aquaculture and rural development.

2.2 The 1970s

The Reserve Bank of India's (RBI) Agricultural Credit Department (ACD) began operations in the late 1970s, while the National Bank for Agriculture and Rural Development (NABARD) was founded in 1982. NABARD was a key player in financing aquaculture projects and rural development programs.

2.3 The 1980s

Expansion of aquaculture: During the 1980s, new technologies and methods such as pond aquaculture, cage culture, and shrimp farming were introduced, leading to considerable expansion of aquaculture.

Government schemes: A number of government programs have been introduced to assist rural aquaculture. These included financial aid for pond construction, inputs, and technical support.

2.4 The 1990s

Policies of liberalization: The 1990s economic liberalization initiatives boosted private sector involvement in aquaculture and rural development. During this time, corporate aquaculture farms began to appear and infrastructural development investments were made.

Microfinance initiatives: Small-scale fish farmers and rural business owners began to receive financial assistance from microfinance institutions.

2.5 The 2000s

Sustainability as a focal point: There was a change in favor of promoting sustainable aquaculture practices as concerns about social and environmental repercussions grew. This includes actions to reduce pollution, better manage water, and improve rural residents' quality of life.

Government support: To assist small-scale aquaculture farmers and promote rural development, the Indian government has implemented a number of programs and subsidies.

2.6 The 2010s

Technological advancements: To boost productivity and efficiency, cutting-edge technologies, including integrated multitrophic aquaculture (IMTA) and recirculating aquaculture systems (RAS), were adopted in the 2010s.

Public-private partnerships: More cooperation between the public and private sectors and non-governmental organizations (NGOs) has resulted in creative financing schemes and initiatives that strengthen rural communities' capacities.

2.7 2020s through 2022s

Digitalization: The use of digital technology such as remote sensing and mobile applications has grown in significance for aquaculture financing and management.

Climate resilience: Through adaptation techniques and risk-reduction plans, efforts have been made to increase the resilience of aquaculture systems to the effects of climate change, such as extreme weather events and sea-level rise.

2.7.1 Financing Needs and Sources of Aquaculture Operations

2.7.1.1 Capital investment

i) Cost of pond construction or leasing
ii) Buying equipment to go along with the land (nets, boats, feeds, pumps, aerators, etc.)
iii) Establishing a hatchery or nursery.
iv) Processing and storage facilities.

2.7.1.2 Costs associated with operations

i) Purchase seeds or fingerlings.
ii) Feed and fertilize.
iii) Labor wages.
iv) Energy and petroleum.
v) Repair and maintenance.
vi) Illness prevention and treatment.
vii) Transportation and marketing.

2.7.1.3 Working capital

i) Managing cash flow gaps between stockings and harvesting
ii) To meet recurring expenses during the production cycle

2.7.1.4 Modernization of Technology

i) Contemporary agricultural methods (e.g., biofloc technology and recirculating aquaculture systems).
ii) Automation and monitoring systems.
iii) Programs for genetic improvement.

2.7.1.5 Research and Development

i) Cultivation of Novel Species or Strains.
ii) Enhancing feeding plans and feed formulations.
iii) Disease control and biosecurity precautions.

2.7.1.6 Training and capacity building

i) Worker and farmer skill development initiatives.
ii) Technical assistance and extension services.

2.8 Socioeconomic impact of aquaculture finance

2.8.1 Economic Growth: With considerable contribution to GDP growth, aquaculture has become a prominent economic industry in India. Millions of people working in fish farming and related industries rely on this industry as a source of income. In India, aquaculture grew at an average annual rate of 5.7% between 2000 and 2018, surpassing the global average growth rate according to a study conducted by the Food and Agriculture Organization (FAO).

2.8.2 Employment Generation: Aquaculture has brought about job prospects in rural areas, especially for laborers without land and small-scale farmers. Fish farming creates jobs for rural communities by requiring labor for tasks such as building ponds, stocking, feeding, and harvesting. According to the National Fisheries Development Board, over 14 million people are employed directly or indirectly through aquaculture in India, according to the National Fisheries Development Board (NFDB).

2.8.3 Poverty Alleviation: By providing rural households with alternative livelihood options and income diversification, aquaculture has significantly contributed to the alleviation of poverty. Marginalized populations can participate in productive activities and increase their socioeconomic standing through small-scale fish farming. Research has indicated that the implementation of aquaculture treatment has resulted in a decrease in poverty by raising household income and enhancing food security.

2.8.4 Food Security: By boosting fish availability and productivity, aquaculture greatly enhances food security, particularly in rural areas, where access to wholesome food is restricted. Fish are a vital part of the diet because they are rich sources of minerals, essential fatty acids, and proteins. By reducing the gap between the supply and demand for fish, aquaculture contributes to increased food security, both nationally and in households.

2.8.5 Infrastructure Development: As roads, electricity, and water supplies are necessary for establishing and maintaining aquaculture operations, the expansion of aquaculture has resulted in the development of infrastructure in rural areas. Construction of infrastructure also aids in the development of rural areas.

2.8.6 Food Security: By boosting fish availability and productivity, aquaculture significantly improves food security, particularly in areas where traditional capture fisheries may decline. This lowers malnutrition and helps rural populations to satisfy their protein needs.

2.8.7 Women's Empowerment: Women are heavily involved in many rural aquaculture systems and help with tasks such as feeding, selling, pond preparation, and seed harvesting. Thus, aquaculture promotes gender equality and empowerment in rural communities.

2.9 Public policies and initiatives for financing aquaculture in India

2.9.1 National Fisheries Development Board (NFDB)

The NFDB offers financial support for a range of aquaculture initiatives including the Blue Revolution.

Provide incentives and subsidies to encourage the development of aquaculture technologies and infrastructure.

2.9.2 Rashtriya Krishi Vikas Yojana (RKVY)

RKVY funds initiatives that improve infrastructure, productivity, and market connections to promote aquaculture growth.

2.9.3 National Bank for Rural and Agricultural Development (NABARD)

NABARD provides loans and credit facilities specially designed for aquaculture projects through a number of programs, including the Fisheries Development Fund.

2.9.4 Pradhan Mantri Matsya Sampada Yojana (PMMSY)

The Pradhan Mantri Matsya Sampada Yojana (PMMSY) was introduced to encourage the development of sustainable fisheries and aquaculture.

Offers financial support for the modernization, expansion, and development of the aquaculture industry infrastructure.

2.9.5 Technology Development and Transfer Programs

These are government-led projects that help institutions and individuals engage in financial innovation and technology transfer while concentrating on research and development to enhance aquaculture technologies.

2.9.6 State-level Initiative

Many Indian states have their own initiatives and plans to assist in financing aquaculture that are adapted to the demands and goals of the region. Subsidies, grants, and credit options are frequently included in these programmes for small and marginal farmers working in aquaculture.

2.9.7 Insurance Schemes

A few insurance programs supported by the government help aquaculture farmers reduce risk by providing monetary security against losses caused by illnesses, natural disasters, or other unfavorable circumstances.

2.9.8 Capacity Building and Training

Programs, workshops, and skill development initiatives are funded to improve the technical and managerial skills of aquaculture practitioners and, consequently, their access to capital and markets.

2.10 Risk Management and Insurance in Aquaculture Financing

2.10.1 Market Risk Assessment

Assess the price volatility and market demand of aquaculture products. Examine rivalry and prospective market disruptions.

We consider variables that affect market dynamics such as customer preferences, legislative changes, and global trade policies. (Arthur, J. R. 2008)

2.10.2 Operational Risk Management

Evaluate operational risks pertaining to farm technology, equipment, and infrastructure. Put precautions to lessen the chance of disease outbreaks, environmental dangers, and natural calamities.

Creating backup plans for emergencies or production hiccups.

2.10.3 Financial Risk Analysis

We carry out financial viability analyses to assess the profitability of aquaculture projects and cash flow forecasts.

Evaluate the risks associated with debt servicing, operating costs, and capital investments.

Use techniques to manage financial risk, such as trading against interest rates and currency volatility.

2.10.4 Regulatory Compliance

Assure compliance with local laws regulating food safety procedures, environmental norms, and aquaculture operations.

To reduce the risk of noncompliance, keep informed of changes in regulatory requirements and modify corporate processes accordingly.

2.10.5 Insurance Protection

Obtain the right insurance to protect against liability risks, property damage, and business interruption.

Collaborate with insurance companies to customize insurance plans for the unique hazards encountered by aquaculture enterprises.

2.10.6 Supply Chain Risk Management

Assess the risks related to supply chains for inputs including feed, seeds, and equipment.

Create backup plans and diversify sources to reduce supply chain interruptions.

Keep an eye on managing the risks associated with distribution channels, logistics, and transportation.

2.10.7 Mitigation of Environmental Risks

Use sustainable aquaculture methods to reduce the negative effects on the environment and regulatory risks; keep an eye on habitat impacts, water quality, and waste management methods to ensure that environmental regulations are being followed; and invest in methods and technologies that save resources and increase productivity.

2.10.8 Evaluation of the Credit Risk

Assess the creditworthiness of counterparties and borrowers in financing deals related to aquaculture.

The risk of default or non-payment is determined based on collateral, credit history, and financial performance.

To track and reduce credit risks during the lending process, establish policies and procedures for credit risk management.(Tietze, U. & Villareal, L. V. 2003)

2.10.9 Risk of Market Entry and Expansion

Evaluate the risks associated with market entry tactics such as the creation of new products, global expansion, and strategic alliances.

The possible risks and returns of expansion projects are evaluated by conducting in-depth market research and feasibility assessments.

Create strategies for gradual expansion and backup plans to mitigate the hazards involved in expanding aquaculture operations.

2.10.10 Monitoring and reporting of risk

Place-in-place mechanisms for continuing performance monitoring, risk assessment, and reporting.

Determine performance metrics and key risk indicators (KRIs) to evaluate the efficacy of risk-management techniques.

Update and review risk management procedures and guidelines on a regular basis in response to evolving risks and shifting market conditions.

3. Future outlook and recommendations

3.1 Technological advancement: Technological developments in aquaculture are likely to lead to higher industrial productivity and efficiency. This includes advancements in fish breeds with improved genetics, precision aquaculture methods, and recirculating aquaculture systems (RAS).

3.2 Market expansion: The aquaculture sector is predicted to develop because of the growing demand for seafood brought on by population growth, rising incomes, and health-conscious customer preferences. This may have resulted in market expansion and increased investment opportunities.

3.3 Sustainable Practices: Aquaculture will see greater emphasis on sustainability due to customer choices and environmental concerns. Long-term success depends on the adoption of sustainable practices, such as waste management, ecosystem-based techniques, and ethical feed procurement.

3.4 Integration with Agriculture: By combining aquaculture (aquaponics) with other industries such as waste management and renewable energy, it may be possible to forge new alliances and spur opportunities for rural development.

3.5 Climate Resilience: Aquaculture faces a number of important issues as a result of climate change, including variations in temperature, acidity of water, and extreme weather. The viability of the aquaculture sector depends on the development of infrastructure and techniques that are climate-resilient.

4. Recommendations

4.1 Access to Finance: Through programs such as microfinance, credit facilities, and government subsidies, small-scale aquaculture producers' access to financing can be increased, particularly in rural areas. (Tietze, U et.al,2007) Both entrepreneurship and rural development could benefit from this.

4.2 Capacity Building: Enhance aquaculture farmers' technical proficiency, understanding of best practices, and business management skills by providing training and capacity-building programs. Consequently, the industry will become more competitive and productive.

4.3 Infrastructure Development: To assist the expansion of the aquaculture sector in rural areas, investments in infrastructure development, such as access roads, electricity, water supplies, and cold storage facilities. As a result, farmers have better market access and less postharvest losses.

4.4 Regulatory Frameworks: Create and implement rules that safeguard food safety, encourage environmentally friendly aquaculture methods, and support sustainable aquaculture operations. This covers certification programs, water quality standards, and zoning laws.

4.5 Research and Innovation: Through funding, collaborations, and knowledge-sharing programs, support the advancement of research in genetics, aquaculture technology, and sustainable practices. This will promote ongoing development and flexibility in response to new difficulties.

4.6 Market Access: Promote aquaculture product entry into the market through trade agreements, market connections, and branding campaigns.

Farmers' livelihoods will improve, and their ability to obtain higher prices for their goods will be enhanced.

Conclusion

In conclusion, by fostering sustainable practices, improving food security, and offering economic opportunities, aquaculture financing is essential for rural development. Rural communities can improve their livelihoods, reduce poverty, and diversify their sources of income by investing in aquaculture projects. Small-scale farmers with access to financing can grow their businesses, implement cutting-edge technologies, and boost production. The construction of ponds, hatcheries, and processing facilities is made possible by financial support, which increases the competitiveness of the aquaculture industry. However, issues such as hard-to-obtain funding, risky investments, and weak regulatory environments still exist and need to be addressed. To overcome these obstacles and realize the full potential of aquaculture for rural development, cooperation between governments, financial institutions, non-governmental organizations, and local people is essential.

References

Arthur, J. R. 2008. General principles of risk analysis process and its application to aquaculture. In M.G. Bondad-Reantaso, T.R. Arthur & R.P. Subasinghe (eds). Understanding and applying risk analysis in aquaculture. FAO Fisheries and Aquaculture Technical Paper No. 496. Rome, FAO. 148 pp

Clement Allan Tisdell et.al 2012, Investment, insurance and risk management for aquaculture development.

FAO. (2014). The State of World Fisheries and Aquaculture Opportunities and challenges. Rome: Food and Agriculture Organization of the United Nations.

FAO. 2022. Promoting sustainable aquaculture for food security and economic development. Harare. https://doi.org/10.4060/cc0324en

https://thedocs.worldbank.org/en/doc/9442516049603965510080022020 original/3BeyondthePondEN.pdf

https://www.fao.org/3/a1182e/a1182e.pdf

https://www.fao.org/3/i6488e/i6488e.pdf

https://www.statista.com/statistics/264577/total-world-fish-production-since-2002/

https://www.statista.com/statistics/621507/rural-and-urban-population-india/

Prein, M. and M. Ahmed, 2000. Integration of aquaculture into smallholder farming systems for improved food security and household nutrition. Food Nutr. Bull. 21: 466-471.

Tietze, U. & Villareal, L. V. 2003. Microfinance in fisheries and aquaculture: guidelines and case studies. FAO Fisheries and Aquaculture Technical Paper No. 440. Rome, FAO. 114 pp.

Tietze, U., Siar, S.V., Marmulla, G. & Van Anrooy, R. 2007. Credit and microfinance needs in inland capture fisheries development and conservation in Asia. FAO Fisheries Technical Paper No. 460. Rome, FAO. 138 pp.

Tveteras, S., Asche, F., Bellemare, M. F., Smith, M. D., Guttormsen, A. G., Lem, A., Vannuccini, S. (2012). Fish Is Food - The FAO's Fish Price Index. Plos One, 7(5), 10. doi:10.1371/journal.pone.0036731.

Colour Plates

Chapter 1: Historical Overview of Aquaculture and Fisheries and Its Evolution

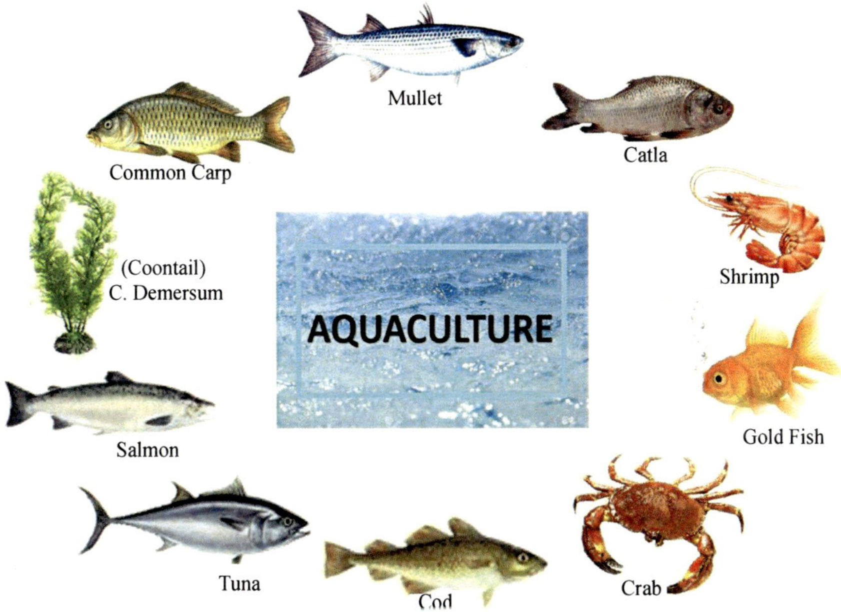

Fig. 1. Image showing different types of organisms cultivated in aquaculture. (Redrawn from- https://www.fisheriesindia.com/2021/04/types-of-aquaculture-and-recent.html)

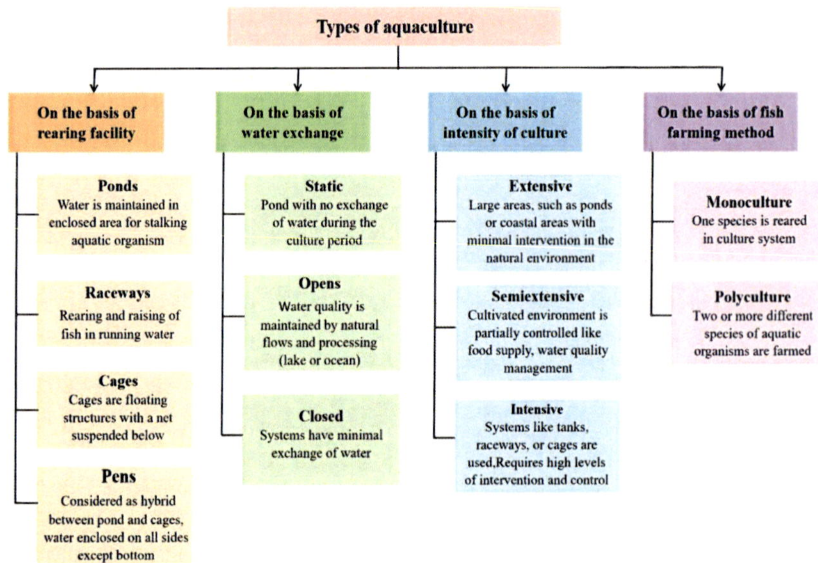

Fig. 2. Image showing types of aquaculture on the basis of rearing facility, water exchange, intensity of culture and fish farming method. (https://www.fisheriesindia.com/2021/04/types-of-aquaculture-and-recent.html)

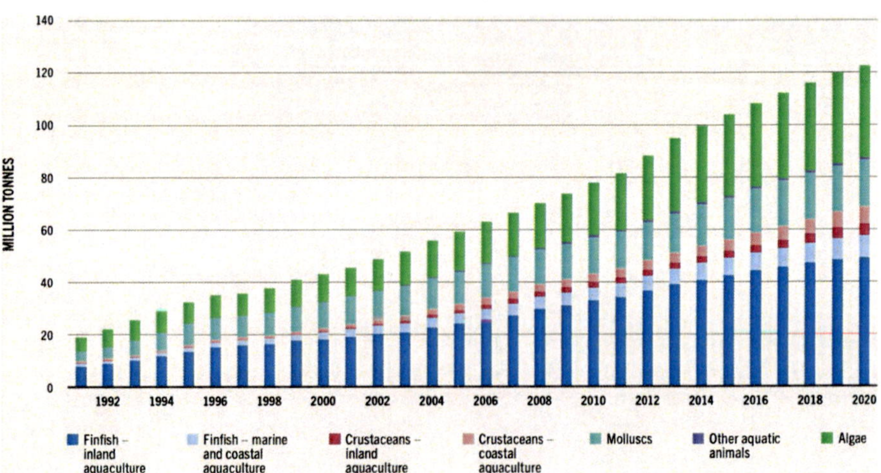

Fig. 4. Global aquaculture production of major species groups and other aquatic animals in million tonnes, 1992–2020, as reported by the FAO. (https://www.fao.org/3/cc0461en/online/sofia/2022/aquaculture-production.html)

Fig. 5. Image showing a Oyster farming (https://www. istockphoto. com/photos/oyster farming), b Seaweed(https://phys.org/news/2017-09-seaweed-fueled-cars-day-tech.html). Fish (https://www.telegraphindia.com/india/bihar-govt-to-seek-gi-tag-for-mithilas-rohu-fish/cid/1859410) Shrimp (https://7esl.com/crustaceans/e. Turtle (https://nas.er.usgs.gov/taxgroup/reptiles/). Frog (https://aquariussystems.blog/)

Fig. 6. Capturing fish in a pond. Engraving from 1582 by Han Bol. (Rijksmuseum, Amsterdam)

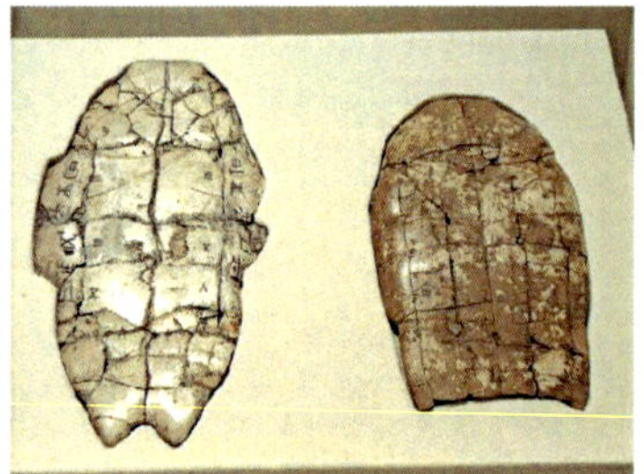

The earliest evidence of aquaculture in China can be found in the marks on ancient "oracle bones." Pieces of turtle shell or bone used in ancient China for divination practices. These inscriptions on oracle bones provide insights into various aspects of ancient Chinese society, including early forms of aquaculture.(Nash *et al.* 2010)

Image a : Oracle bones on display at the National Museum of China in Beijing, March 10, 2023.[Photo/CGTN]

The specific animals such as crocodile, Ibis, Dog-Headed Ape, and Fish are mentioned as being venerated in ancient Egypt.The Egyptians began to keep fishes in sanctuaries and worship them as deities incarnate. This interpretation is based on historical and archaeological studies of ancient Egyptian beliefs and practices.(Wallis *Budge et al.* 1904)

Image b : Central Garden Pool in the Garden of Nebamun's Tomb Painting, British Museum, late 18th Dynasty, circa 1350 BCE (Photo in public domain)- ELECTRUM MAGAZINE

Fig. 7. Showing the facts which indicates the religious importance of aquaculture in ancient time

Fig. 8. Images showing some examples of koi that display variable patterns of scale color.

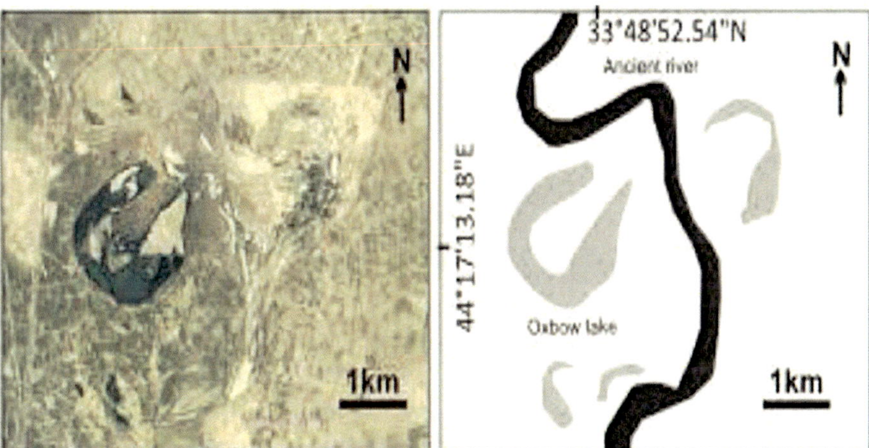

Fig. 9. Image showing oxbow theory of aquaculture. Sources-https://www.researchgate.net/publication/329718098_Recognition_criteria_for_canals_and_rivers_in_the_Mesopotamian_floodplain/figures?lo=1

Fig. 10. Image showing pond made to defend Castle Sources-https://www.shutterstock.com/image-photo/caerlaverock-castle-moated-triangular-first-built-2252331277

Fig. 12. Images of the fishes mentioned in Table 1

a

Azolla Pinnata

b

Spirodela polyrhiza

5 mm

Spirodela Polyrriza

c

Lemna minor

2 mm

Lemna Minar

d

Woiffia Arrhiza

Fig. 13. A. Azolla-https://idtools.org/fnwd/index.cfm?packageID=1097&entityID=2558, b. Spirodela by Ulrich Kutschera and Karl J Nikalas, 2014, c. Lemna by Ulrich Kutschera and Karl J Nikalas,2014, d. Wolffia- https://www.knowyourweeds.com/hi/weeds/Wolffia_arrhiza

Fig. 15. Image of integrated fish farming (https://www.asiafarming.com/integrated-farming-fish-livestock)

Fig. 16. Pen culture (https://aquafind.com/articles/Modified_Aquaculture_Systems_In_India.php)

Fig. 17. Cage culture (five cages and two cages under All India Network Project on Mariculture of ICAR- Central Marine Fisheries Research Institute (CMFRI), Karwar, India.)

Chapter 3: The Future of Aquaculture: Prospects and Challenges

Fig. 1. Farm showing Aquaculture Setup.

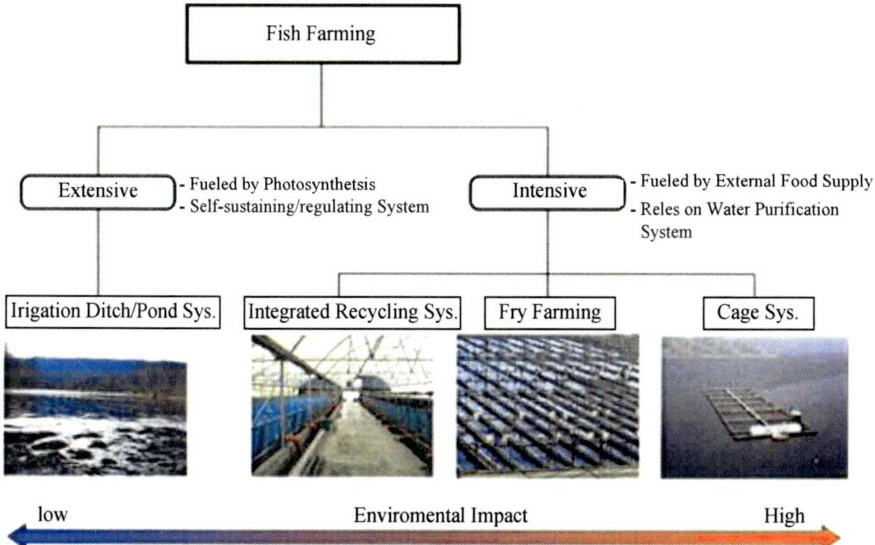

Fig. 2. Picture showing different types of aquaculture.

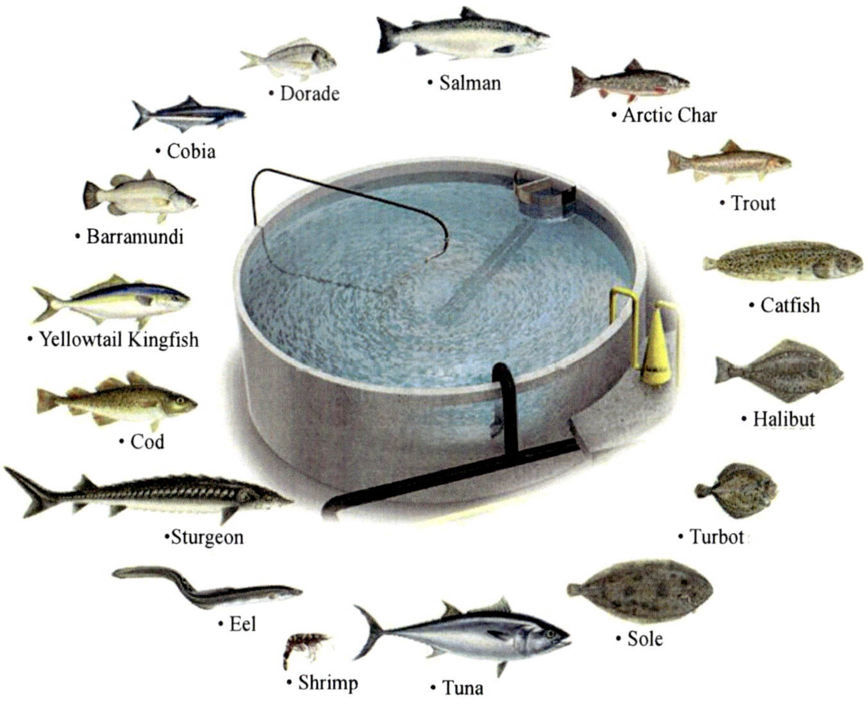

Fig. 3. Different types of fish used in Aquaculture.

Chapter 5: Aquaculture and Its Impact on Environment

Fig. 1. Aerial view of a mariculture farm in a bay. (The Pearl Protectors, 2023)

Integrated Multi-Trophic Aquaculture
(Imta)

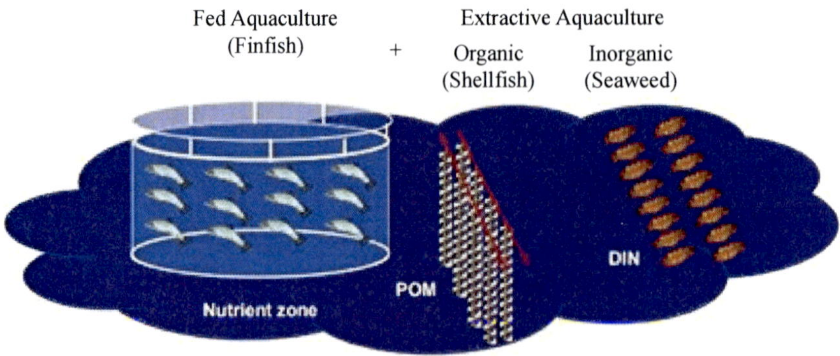

Fig. 2. Conceptual diagram of Integrated Multi-Trophic Aquaculture (IMTA) (Soto, 2009)

Fig. 3. Three circular fish cages float in a lake (Soto, 2009)

Chapter 7: Minimal Water Usage System in Aquaculture

Aquaponics System

Grow Tray

Clean Water return

Air line

Fish Tank

Nutrition Pump

Air Stone

Air Pump

Fig. 3. Aquaponic system

Chapter 8: Precision Fisheries: The Future of Farming

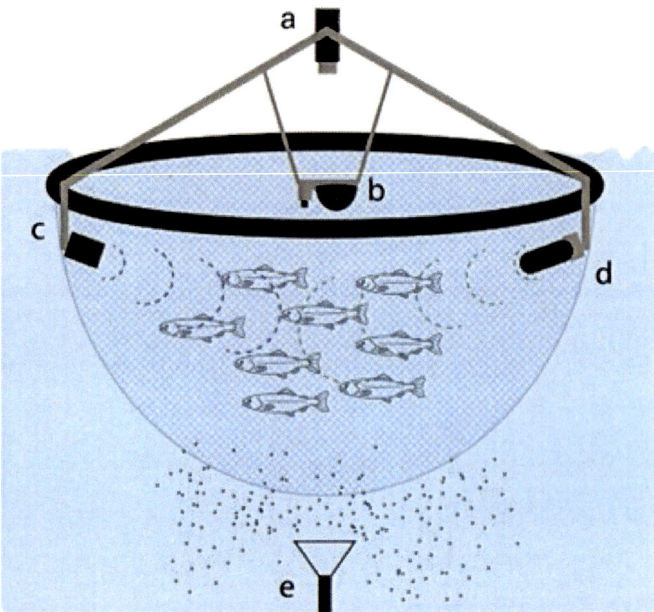

Fig. 1. Example of precision fish farming system including (a) surface camera, (b) underwater camera with multiparameter water probe, (c) sonar, (d) acoustic telemetry system with hydrophone, and (e) smart trap (Braña *et al.*, 2021).

Chapter 9: Sustainable Aquaculture Practices and Approaches

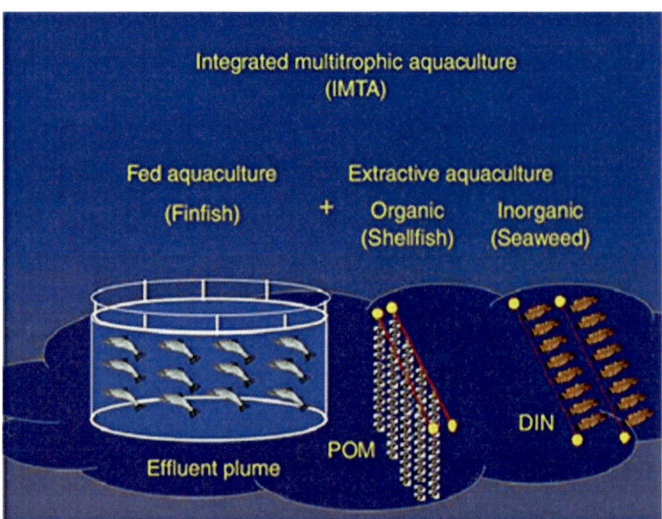

Fig. 1. Conceptual diagram of an integrated multitrophic aquaculture (IMTA) (Chopin *et al.*, 2008)

Chapter 10: Recent Trends and Innovative Technology in Fishing Technology

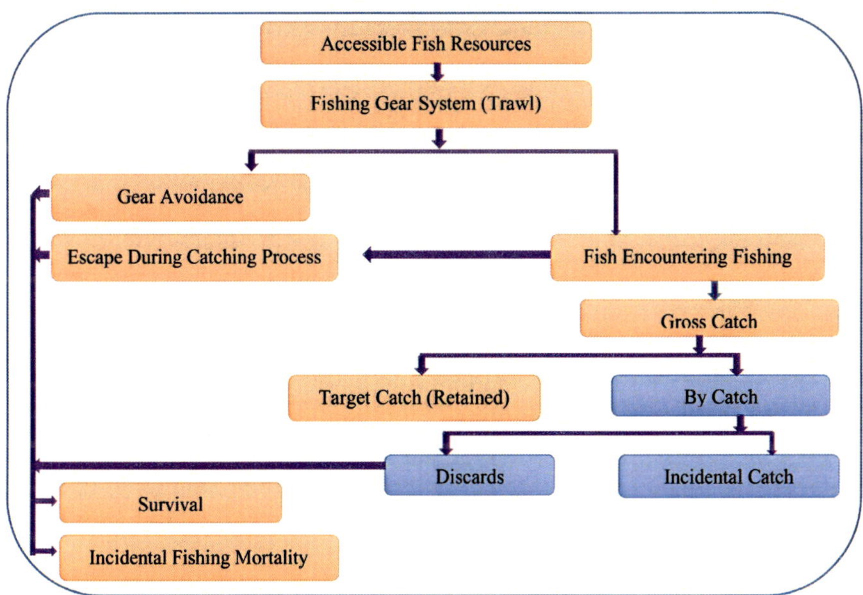

Fig. 1. Catch process and bycatch production in trawling

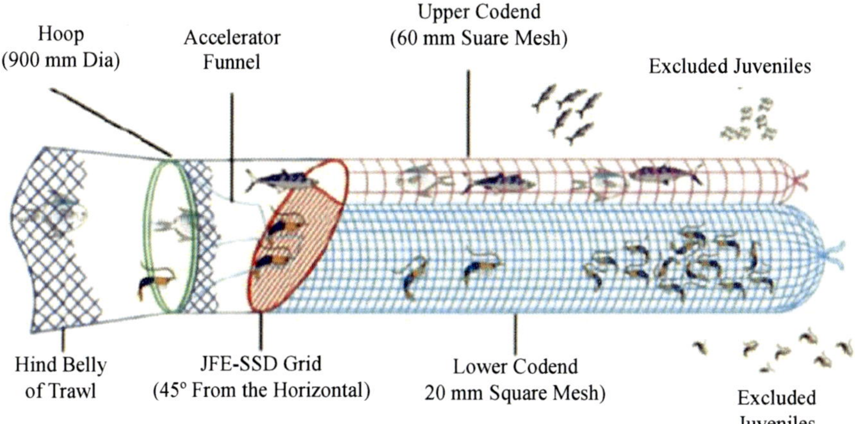

Fig. 7. Schematic diagram of JFE-SSD

Chapter 12: Optimizing Nutritional Quality in Fried Fish Through Artificial Intelligence Innovation

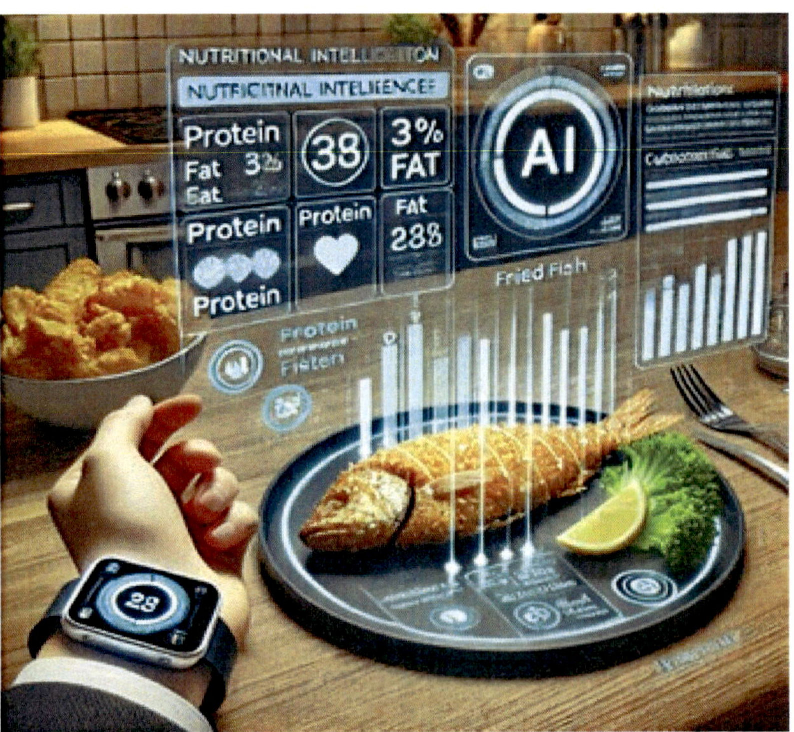

Fig. 1. Showing the Depicting fried fish being analyzed through artificial intelligence, showcasing nutritional data in a futuristic setting.

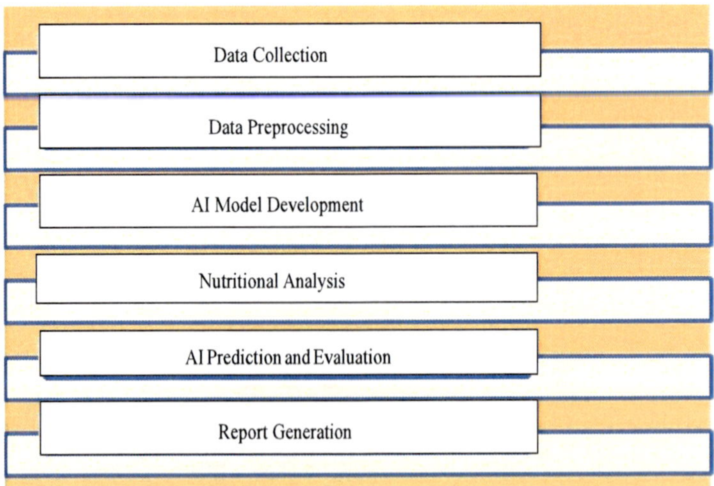

Fig. 2. Shows the Nutritional Value of Fried Fish Using AI Approach

Chapter 13: Integrated Fish Farming as Future Importance and Increasing Possibilities of Income to Fish Farmers

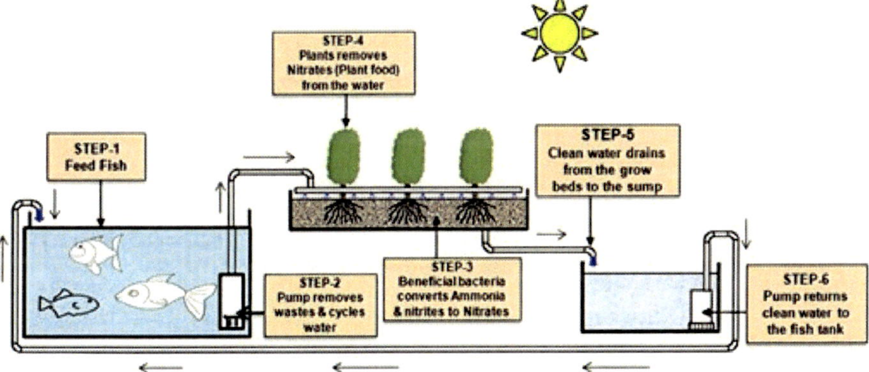

Fig. 1. Aquaponics system In Aquaponics system (Jena *et al.*, 2017).

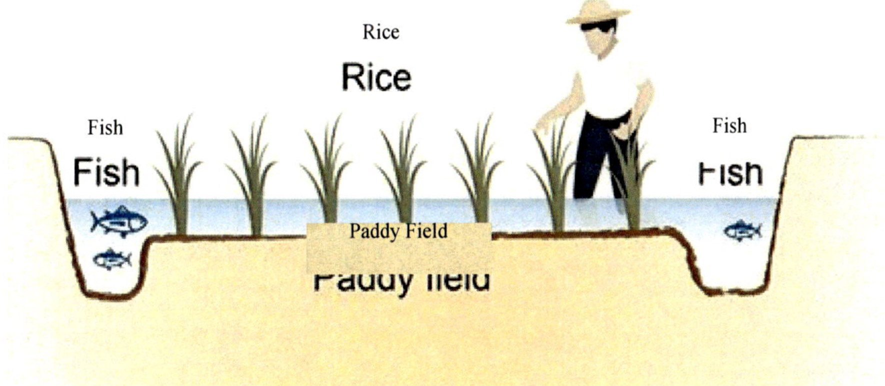

Fig. 2. Rice-Fish Culture (Yi, 2019)

Chapter 15: Organic Fish Farming: Certification for Sustainable Income and Future Importance

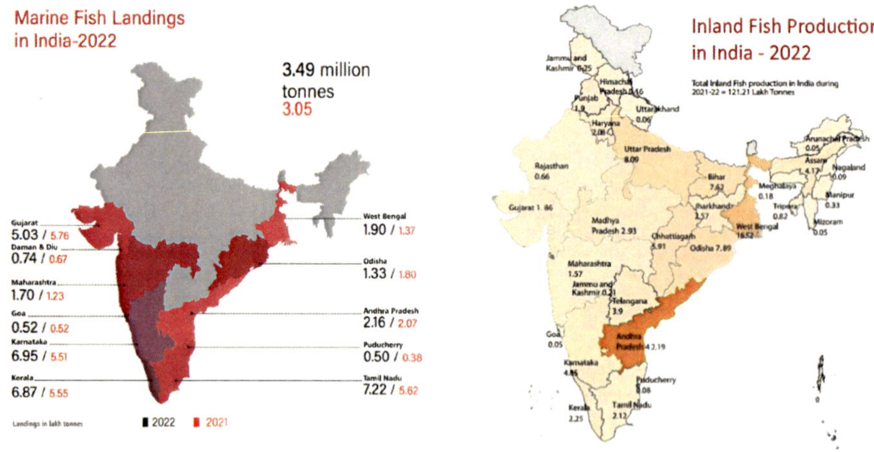

Fig. Fish production in India in the year of 2022.

Source: CMFRI Annual Report 2022, Handbook of Statistics on Indian State -RBI (2021)

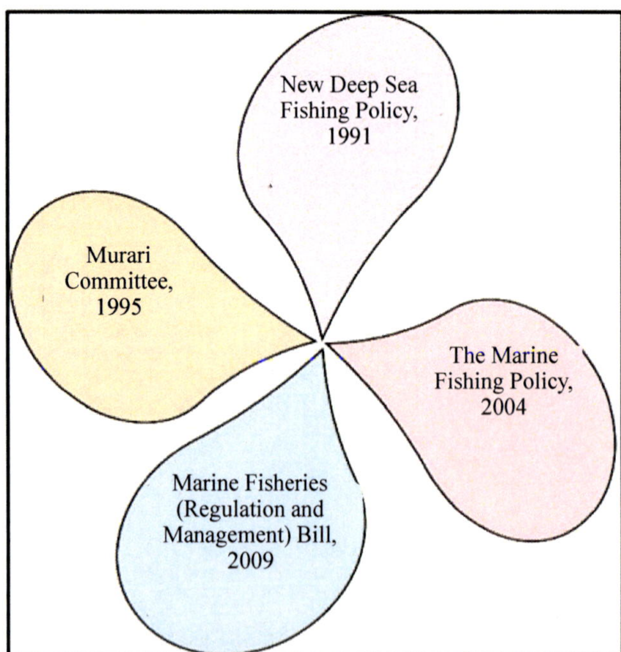

Fig. 1. The Indian fisheries policies and committees